Hermam Vargas Silva

Environmental performance assessment for small mining operations

Hermam Vargas Silva

Environmental performance assessment for small mining operations

Proposed audit criteria

ScienciaScripts

Cover image: www.ingimage.com

This book is a translation from the original published under ISBN 978-3-330-75797-4.

Publisher:
Sciencia Scripts
is a trademark of
Dodo Books Indian Ocean Ltd. and OmniScriptum S.R.L publishing group

120 High Road, East Finchley, London, N2 9ED, United Kingdom
Str. Armeneasca 28/1, office 1, Chisinau MD-2012, Republic of Moldova, Europe
Printed at: see last page
ISBN: 978-620-8-29274-4

SUMMARY

Thanks

To the Federal University of Santa Catarina, for accepting me into the PPGEP. To my advisor, Prof. Gregório Varvakis, for his conversations, patience and, above all, competent supervision.

To the course lecturers for demonstrating their competence and innovation in dealing with environmental and educational issues.

The director of the Ilha Grande National Park, lawyer Maude Joslin Mota, for the opportunity and trust.

To my fellow students for the good and bad times when we learned a little more about life.

The Rectory of UNIPAR - Universidade Paranaense, for the opportunity and trust placed in me.

SUMMARY

SILVA, Hermam Vargas, Uma proposta de indicadores de desempenho ambiental para extrações de areia, dissertation presented to the Postgraduate Program in Production Engineering - PPGEP, of the Federal University of Santa Catarina - UFSC, 2002.

This dissertation is the result of the demand for environmental performance information for sand extraction by dredging. It aims to combine the development of a system of environmental indicators with a management practice that is more in line with modern times. Its general objective is to present a set of environmental performance assessment indicators for sand mining by dredging, drawing up guidelines that should be answered and/or observed, making up a report or environmental report of the situation. The working method adopted was the description of the characteristics of a given phenomenon or the establishment of relationships between variables. Documents, bibliographies and reported experiences were used as elements of the study, validated in two extraction sites in the environmental protection areas, APA, of the Intermunicipal Consortium for the Preservation of the Remnants of the Paranâ River. The work presents a set of indicators in eight (8) areas: environmental policy; environmental management system; risk management system; legal compliance; product management; financial performance, environmental development and accounting criteria, which are presented in the form of items and discussed. The set of environmental performance indicators applied reports on the environmental situation of the companies and appears to be adequate for what was proposed. In the end, it establishes a hierarchy of environmental behaviour in 4 (four) classes (reactive, in the beginning of implementation, in implementation and proactive), with the two mining companies studied being in the first class (reactive).

Chapter 1 - Initial Considerations

1.1 Research justification

This research is the result of the demand for environmental performance information to help make business decisions.

It wants to combine research into new techniques (e.g. environmental performance assessment) with the development of a system of environmental indicators, with the aim of communicating externally the results of the Environmental Performance Assessment (EPA) of riverbed sand extraction companies.

With this in mind, the dissertation focused on analyzing the indicators that could form such an assessment, with the aim of discussing a proposal for an ADA for sand mining operations that exploit minerals by dredging in riverbeds.

The importance of this work is justified by the fact that bibliographical research has failed to detect this type of environmental performance assessment for sand extraction areas by dredging, although various types of ADA were found for companies.

This application aims to contribute to its use in environmental monitoring of sand mining, trying to develop criteria that, if applied, can at least protect the mining company from being classified as a defendant in the event of an environmental accident, since mining, especially those ores classified as type II (those directly used in civil construction without processing) under the Mining Code, has a special character in Brazil's environmental policy.

The discussion of indicators will help the municipal, state and federal environmental agencies with regard to their purpose of overseeing mining, and their correct use could be a good indicator for license review attitudes, for maintaining or eliminating sand mining activities, by dredging and in riverbeds.

1.2 General objective

Develop a set of ADA indicators for riverbed sand dredging areas, drawing up a set of guidelines that must be observed and responded to, forming a report or situation report.

1.3 Specific objectives

- Evaluate the conceptual basis for ADA indicators;
- To provide a comprehensive overview of the impacts caused by sand extraction;

- Create ADA indicators for sand mining by dredging in riverbeds;

- Discuss these indicators for riverbed sand mining by dredging.

1.4 Considerations about the working methodology for this dissertation

The proposal to discuss a group of ADA indicators for sand extraction by dredging in riverbeds is based on the ideas of GIL (1995:44-46), who describes "descriptive research" as that which has as its main objective the description of the characteristics of a given phenomenon or the establishment of relationships between variables, some of which go beyond simply identifying the existence of relationships, aiming to determine the nature of this relationship. The main elements of the study were documents, bibliography and experiences gathered over the years from work in the field of small-scale mining, in conjunction with environmental issues.

In this sense, the first phase of data collection had as its main sources of information:

- the academic theoretical production on performance evaluation methods in companies, providing a strong theoretical vision that facilitated the understanding of the paths proposed by the study;

- the search for documentary records on Environmental Performance Assessment in companies and/or *papers* related to the subject;

- collection of the experiment, in terms of environmental impact assessment reports, reports on the recovery of degraded areas and visits to the environmental agency (Instituto Ambiental do Paranà - IAP);

- a stage of compiling this data into a table that would meet the objectives of this dissertation and that of validation;

- and finally, validation in two sand mines, by dredging and in a riverbed.

In short, the steps that will be taken in this dissertation can be characterized by :

a) report on the ADA models found;

b) Establishment of an Environmental Impact Assessment (EIA) for sand mining using the dredging method, in a riverbed, as the initial state for analysis;

c) relate the data obtained by the EIA to the ADA models;

d) discussion of the results;

e) presentation of the indicators and their validation.

1.5 The structure of the work

Chapter 1 outlines the construction of the work, its general and secondary objectives and the methodology.

Chapter 2 presents the theoretical background to the approaches to environmental issues and the company, performance evaluation and environmental performance evaluation.

Chapter 3 discusses the indicators used to evaluate the environmental performance of dredged and riverbed sand mines and proposes a set of indicators to evaluate the performance of dredged sand mines.

Chapter 4 validates the set of indicators obtained, discusses the data and offers some considerations.

This is followed by Chapter 5, which concludes the work and then references the sources of information cited.

Chapter 2 - The environmental issue and performance evaluation

2. 1 Introduction

Man has always used the resources that nature provides, generating "utilities" and "waste" with this transformation, which are disposed of again in the natural environment without any major concern.

MOURA (2002), highlights three periods for the development of environmental issues in the world, the first in the 1960s, with the increase in population and consumption, the occurrence of some major accidents and the forecast of a future depletion of some important resources (oil, wood, water, among others), provided some repercussions, such as the "Meadows Report", using projections for population growth, pollution and use of water, among others, wood, water, among others), led to some repercussions, such as the "Meadows Report", which, using projections for population growth, pollution and the use of the earth's natural resources, predicted their depletion. Also in this period, Rachel Carson published "Silent Spring", apud MOURA, op. cit., which explored the theme of a spring without birds due to the use of DDT, which caused great debate.

In the same decade, England began a process to decontaminate the River Thames and intense efforts were made to improve London's air quality, according to PARIZOTO (1995); in the USA, movements for consumer protection, promotion of minorities and women, product quality and safety, among others, appeared, which noted the effects of pollution on people's health.

A second period would comprise the 1970s and 1980s (MOURA, op. cit.), where environmental contamination caused serious repercussions, such as Minamata Bay in Japan, where the surrounding companies dumped their effluents containing heavy metals, which were absorbed by the local food chain and contaminated the population, resulting in around 2,200 victims, officially recognized by the Japanese government. Legislative activity also increased, which, together with greater organization of social movements, forced the creation of laws and better environmental control techniques. In 1972, the United Nations Conference on Environment and Development took place in Stockholm, CASTRO (1996), which dealt with the discrepancy between rich and poor, developed and underdeveloped countries, compounded by an energy crisis caused by the OPEC countries (Organization of Petroleum

Exporting Countries), an alarming situation, SEBRAE (1996).

The basic concern was with the increase in population and the rise in pollution levels and the finiteness of natural resources.

In 1987, the World Commission on the Environment produced the concept of "sustainable development", which allows humanity to use natural resources without compromising their use by future generations. The "Blue Angel" ecolabel (Blau Engel) appeared in Germany in the 1980s to certify environmentally friendly products; in the same period, the USA began to require Environmental Impact Assessments (EIA), with public hearings and approval at different levels of government organizations, as a prerequisite for approving polluting projects, creating a series of laws (Clean Air Act, Clean Water Act, Resource Recovery and Conservation Act, Comprehensive Environmental Responsibility, Cleanup and Accountability Act, the *Superfund*, Safe Drinking Water Act, Toxic Substances Control Act, Mining and Land Reclamation Act and Community Right to Information Act) as SANCHES (2001) informs us.

Still in the 1980s, there were a number of accidents that had a major impact, among them, OSTRONOF (1993) cites the Exxon Valdez oil tanker in Alaska; the Union Carbide plant in Bhopal, India; the tetrachloro dibenzene dioxin (TCDD) leak in Seveso, Italy; the atomic power plant in Chernobyl, in the then Soviet Union; and the Love Canal allotment in the United States. In 1987, the World Commission on Environment and Development issued a report entitled "Our Common Future", or the "Brundtland Report", which revived the discussion of sustainable development, making it an international concept ALMEIDA, MELLO and CAVALCANTI (2000).

And finally, there was a third period, the 1990s, when the big discussions on environmental issues really began. The major event was the United Nations Conference on Environment and Development in Rio de Janeiro, known as the Earth Summit, Rio 92 or Eco 92. This conference showed that a serious and profound change was underway in the understanding of global environmental issues, associated with a need for development, especially by third world countries (ALMEIDA, MELLO and CAVALCANTI, op. cit).

The documents produced were Agenda 21, the Rio Declaration on Environment and Development and the Climate and Biodiversity Conventions. As can be seen, the discussions at ECO-92 are a process of re-reading the economy, not just taking into account cause-effect processes, but a complex whole, which includes political and public administrative aspects, as well as socio-economic discussions, in a binomial of growth and economic well-being, with preservation and conservation of nature, "ecodevelopment" JUCHEM (1992).

2.2 The question of performance evaluation

In this last period (the 1990s), cited by MOURA (op. cit.), companies began to understand that the environment was part of their activities and, seeing the strategic nature of meeting new consumer profiles, they approached the environmental issue from the point of view of ecologically correct products, SIMM (2000). In the same way, organizations also felt that these practices and procedures reduced costs in industrial processes, SEBRAE (1996). In this way, measures (variables) or indicators (ratios between variables) have been created for company performance, initially with a view to quality and process, and then looking at other issues. Performance measures are the "vital signs" of the organization, according to HRONEC (1994), because they quantify how the activities of a process achieve their specific goals.

These indicators guide people, indicate whether they are acting as part of the company's larger strategy. They detect and control improvement within the process, communicate process results from lower to higher levels and create what is important to the whole organization.

2.2.1 Benefits of performance measures

From an integrated perspective, customer satisfaction (internal/external) is sought, where performance measures reflect concerns about the industrial process, thus giving rise to others related to environmental goals, which can consequently lead to improvements in profitability and the company's image.

As continuous improvement is the key to long-term performance HRONEC (1994), performance measures make process improvement not only possible, but also continuous, with the indicator tracking the number of activities over time being a measure that encourages managers and/or employees to progressively reduce the number of phases or actions in the process in order to accelerate speed, flexibility and organizational creativity.

As *benchmarking* is a structured methodology for measuring and evaluating processes and products against others, it is also considered one of the best practices for indicating the best way to execute a process.

The right performance measures help organizations to change successfully, because they define and reward the new behavior, aligning the company's activities with the organization's goals, ANDRADE, TACHIZAWA and CARVALHO, (2000).

The practice of organizational management has shown that the benefit obtained from change is an effective, less expensive way of transforming human behaviour, through evaluations of its environmental performance and also a way for the company to guarantee its existence in the market, DEMAJOROVIC and SANCHES (1999).

In order to do this, the company or organization needs to have clearly defined what the measures are for the indices that its customers expect and what their relative importance is, from various points of view, for example, effectiveness (are we doing the right thing?), efficiency (are we doing it the right way?), quality (is it suitable for use?), quality of life at work (the internal environment, including the work and capital consumed).), innovation (is it capable of generating or creating value?), productivity (the relationship between the value added by the process versus the value of the work and capital consumed), quality of life at work (internal environment, including not only occupational safety legislation) and profitability (for companies) or budgetability (for non-profit organizations), RADOS (2000).

Performance indices need to be understood systematically as a set of measures that make it possible to visualize, analyze and improve the factors for the success of the organization CARO (1998); on the other hand, they also incorporate measures of less objective dimensions (customer sentiment, teaching quality, among others).

These measures need to be expressed by a number and a metric system, indicating respectively their magnitude, "how much", and their meaning, "what". Often multidimensional indicators are used, such as proportions or ratios of two or more fundamental units (kilometer per liter, hours worked per number of accidents), which should be expressed in units that can be better understood by those who will benefit from them for decision-making.

2.2.2 Systems for measuring performance

This data collection work was based mainly on the nomenclature presented by CARO (1998) regarding various approaches, as shown in the table below.

Table 1 - The advantages and disadvantages of the families of performance measures.

APPROACH	ADVANTAGES	DISADVANTAGES
MOREIRA	Support for organizational objectives and strategies.	• No defined tools or techniques for its operation; • Lack of process vision
FAMILY NEVADA QUALITY FORUM	Sequence for defining process measures	- It does not emphasize the need for alignment with tactical and strategic results.
SANDIANATIONAL LABORATORIES	Structure and procedures based on a systemic concept; Introduction to process vision and process results;	The tactical level needs to be aligned with the strategic level;
UNIVERSITY CALIFORNIA	Simple model, suited to the principles or fundamentals of performance measurement	It does not emphasize the customer-supplier view at the process level (horizontal view);
APPROACH	ADVANTAGES	DISADVANTAGES
ERNST & YOUNG	Structured model, defines measures for all organizational levels.	It does not takes into account the interaction between financial and non-financial measures.
JURAN	Concept of *Feedback.*	Operational vision (quality control).
FIELDS	Easily applicable for day-to-day (short-term) work control.	Long-term measures need to be defined (strategic vision).
HRONEC	It offers a comprehensive overview of how to implement a performance measurement system.	It requires a significant commitment of resources and time.

<table>
<tr><td>HARRINGTON</td><td>Defines measures within a process improvement process (simultaneous work) saving effort and money.
It allows progressive adaptation to the process of change and improvement.</td><td>Focus on defining measures at the process level, leaving the definition of organizational goals as the final stage.</td></tr>
<tr><th>APPROACH</th><th>ADVANTAGES</th><th>DISADVANTAGES</th></tr>
<tr><td>RUMMLER & BRACHE</td><td>It interconnects the three organizational levels;
It considers the need to align measures with organizational objectives.</td><td>Performance improvement through engineering; Applicable in dynamic organizations to adapt to change.</td></tr>
<tr><td>SINK & TUTTLE</td><td>Well-defined and well-structured criteria for measuring performance.</td><td>Complexity and high resource requirements; Suitable for organizations with available resources.</td></tr>
<tr><td>KAPLAN & NORTON</td><td>Generates a "unity of purpose"
Measures are aimed at the unified strategy Complements financial measures with non-financial ones.
Operational flexibility.</td><td>without defined tools or techniques for its operation.</td></tr>
</table>

Source : Caro, (1998).

CARO (op. cit.), finally, states that the KAPLAN and NORTON (1997) approach has the greatest advantages, as it allows for the generation of financial and non-financial measures, translating the company's strategy into these perspectives already described, incorporating the customer and their requirements, process analysis, probable innovations and organizational learning, guaranteeing the visualization of the organization and its surroundings and its competitiveness.

They make choices more flexible according to the specific needs and realities of the business and the organization, translating into strategies that add value to the product, PORTO (2001).

SILVA, V. C. also adds (2000), that the use of indicators (or observations), derived from parameters or trends in the state of the environment, becomes a tool for planning the use and evaluation of the situation, evolution and quantification of aspects of reality and, CARVALHO (2000), only recognizes an Environmental Management System (EMS) as complete, if there is the integration of the dimensions of health and safety, together with the analysis of processes, means and management criteria.

Therefore, following these recommendations, we will characterize some ways of evaluating environmental performance.

2.3 Environmental performance assessment

In the discussion on the development of environmental performance indicators, WINSSEMIUS (1997) discusses the subject by distinguishing between four categories of companies, as follows :

- functionally oriented companies, which do not deal with environmental issues; .
- the functionally oriented, who have some experience in dealing with the environment;
- those that are proposing to adopt integrated performance, although they lack technical experience; in addition, they do not have the infrastructure to implement the solutions;
- those that, having learned the lessons, take on environmental challenges in a new way, accepting responsibility for their products, establishing new partnerships with suppliers, consumers and competitors (waste selection and recycling, green labels, agreements and protocols with government agencies).

In the short term, their desire is to stay out of trouble, comply with regulations and laws, prevent environmental accidents, as well as being excessively concerned with waste disposal, forgetting about the big problems. These are understandings that do not have a long-term perspective, leading them to take timid steps.

These companies lack the skills and credibility to define strategic issues, redefine products, recycle, return products quickly, cooperate in agreements or negotiations with commercial partners, in other words, a whole system of business information that no longer expresses the breadth of vision needed to deal with current issues.

Fearful of managers, who would be the people tasked with transforming the strategic vision into operational action, reflected in their responsibility and performance, they tend to be more concerned with economic performance than their superiors. They are professionals who are deeply skeptical about their management's commitment to environmental issues.

In this context, "environmental performance assessments" are objective documents with very different purposes, methodology, procedures and audiences than an environmental audit JUCHEM (1995), because companies are still unsure about the amount of information they should include, the quality of this data, how demanding it is and what kind of audience the report should be aimed at, and whether it should be mandatory or remain a voluntary action.

KINLAW's (1997) proposal is to measure the level at which the company responds to problems that arise and positions itself to remain competitive, assuming four distinct levels of response to environmental pressures:

compliance with the law: this is the first stage of the challenge and is largely linked to meeting legal requirements. The company is already aware of the pressures created

the multiple needs of its stakeholders, awakened citizenship, green consumers and the threat posed by competitors;

non-integrated initiatives: this is where companies begin to go beyond what is required by law, creating and implementing various types of environmental programs, anticipating areas of future risk, reducing their use of energy, reducing the use of packaging and the generation of waste, making it an opportunity for new business. It is a phase of anticipation of the development of legislation, but there is no unifying policy or integrated business objectives;
integrated environmental plans and initiatives: This is where the importance of creating an environmental policy is recognized, with specific objectives for continuous improvement, with a focus on monitoring environmental performance, implementation of training already completed, improvement projects introduced, investments made to develop and/or acquire new environmental technologies, auditing and reporting systems being implemented, participation in associations and management and human resources systems reformulated to support this;

sustainable performance: the company already has a specific policy and targets and the full involvement of its stakeholders, it also has a plan for reviewing this policy and the means to better define new initiatives. It keeps up-to-date information on all input and output processes, waste recovery and secondary products; it also has a well-developed training program, creating specific projects to develop or review all services and products for compatibility between production, nature and its surroundings, supporting research into new technologies to improve its environmental performance;

It also has a fully developed and integrated auditing system to monitor and evaluate its performance. It participates in associations to share information and work together to solve problems.

problems, developing standards and supporting the development of technologies.

The company makes the environment an integral part of all its management systems, with reports to management.

The company's performance is shown in a series of *checklists* (see Table 2):

Table 2 Sustainable Performance Assessment

		Level	of	answer		
MARCO	Answer Zero	Compliance with the law	No integrated	Integrated	Sustainable performance	Classification
a) management demonstrates support						
b) made with the full involvement of the workforce						
c) done with full involvement the parties interested						
		Level	of	answer		
d) politics						

completely compatible comos principles of DS						
e) leaves no doubt that the development values the environment						
f) binds the interests of company to environmental interests						
g) includes specifically the concept of sustainability						
TOTAL						

Source: Kinlaw (1997)

Checklists are also used as a form of eco-audit in the words of CALLENBACH (1999), to assess the corporate culture[1] , trying to understand the company as a metabolic flow of energy (input flows);

energy and materials), projects with eco-friendly criteria (design, processing and manufacturing), output flows (sales, marketing, waste and emissions) and finance (finance, human resources and other support structures).

The ISO 14.000 series of standards provides tools and establishes a standard for environmental management systems, continually improving their indices, D'AVIGNON (1995). One of the most discussed and in-depth areas is standard 14.031 (Environmental

1 It is a set of ideas, values, norms and modes of conduct adopted by a company through consensus. It constitutes the distinctive and unmistakable character of the organization (Gazeta Mercantil, 1996).

Performance Assessment), which is defined as the results obtained from the management of the organization's environmental aspects, REIS (1995), i.e. all actions that may result in the management of the company's activities, products and services, which may interact with the environment. It measures, analyzes, evaluates, reports and communicates, against a certain criterion agreed by the company's management.

In this way, environmental performance is the effective environmental result presented by the company, in the words of ALMEIDA, MELLO and CAVALCANTE (2000), as a function of the compliance[2] achieved by this company and the way in which it ensures this compliance; It will be called "legal" if it complies with the provisions of current legislation applicable to it, but it can also rise to more restrictive and differential parameters, the "corporate" ones, which are additional self-established environmental quality targets, a current trend as a market differentiating factor, especially in the export market.

In order to obtain these numbers or indices, it is necessary to define the impact of the project on the environment, which implies establishing a database so that it can be received, analyzed and interpreted, MULLER, (1995), and these factors can be classified into: fixed (hydrological, topographical and meteorological conditions, among others) which can be modified on a large scale by human action, and variable (location, scale, anthropogenic actions, timing, reactions of the environment, among others) which can also be altered by our decisions or by the pressures they cause on our civilization.

To measure this in the environment, qualitative and/or quantitative metrics are used; in the former, we survey the environment and review the project's characteristics, identifying the sensitive and critical areas to the planned action, qualifying the results of the action, such as the formation of degraded sites, loss of forests and animal species in disagreement with the law, changes in the profile of agricultural occupation, mining and/or interference in urban systems, among others.

Along the same lines, the quantitative form measures values and indices (variations in measurements) of the elements that are part of the analyzed environment. These are values on the scale of the impacts on the factors, which have already been qualified, making it possible, depending on the situation and availability, to anticipate actions on how much and

2 It is the fulfillment of a specified requirement (Gazeta Mercantil, 1996).

which values will change as a result of the implementation of the desired project.

The most significant environmental aspects[3] should be chosen, and at this stage the "process management" methodology can be used as a way of assessing the significant impacts in the main areas, as SILVA and RADOS (2000) point out. Based on this, a selection of indicators can be drawn up for the three areas, namely management, operational and environmental.

VITERBO JÛNIOR (1998) further defines that the indices are those capable of translating the organization's efforts and capacity to manage training, legal requirements, resource allocation, documentation and the application of corrective actions when necessary. For the operational area, the materials, resources and energy used in the production of goods and services. They must also provide information on the consumption of goods and services, resources and energy (entering the process) and the products, atmospheric emissions, effluents, waste, noise, heat, odors, vibration, light and radiation (leaving the production process).

In the environmental field, it is defined as air, water, soil, plants, animal life, other natural resources (e.g. landscapes) and human health.

There are four stages to performance evaluation (data collection, data analysis, information evaluation, reporting and communication of environmental performance).

Below is an example of indicators, drawn up by CAVALIERI, (1997).

Table 3 Examples of environmental indicators

Indicators for the management area	Indicators for the operational area	Indicators for the environmental area
Implementation of policies and <u>programs:</u> number of objectives and targets achieved; number of units in the organization that have	<u>Materials:</u> quantity of materials used per product; amount of packaging material used or discarded per product;	<u>Air:</u> concentration of contaminants in the air; noise; odors;

3 They are defined as products or services resulting from the organization's activity that can interact with the environment (Gazeta Mercantil, 1996).

achieved environmental objectives and targets; number of pollution prevention initiatives implemented; number of employees trained;	quantity of raw materials reused in production;	
Compliance: degrees of compliance with environmental standards; response time to environmental accidents; number of audits implementedversus planned; number of fines received;	Energy: number of units of energy spent per year per unit of product; amount of each type of energy used;	Water: concentracâode contaminants in water; number of coliforms per liter of water; dissolved oxygen;
Indicators for the management area	Indicators for the operational area	Indicators for the environmental area
Financial performance: cost of activities related to environmental performance; research and development funds applied to environmental projects; return on investments related to environmental projects;	Equipment and logistics: average consumption of the vehicle fleet; number of business trips avoided through other means of communication;	Soil: erosion; concentration of contaminants in the soil;
Relationship with a	Products: number of new products	Flora: quality of vegetation;

community: number of environmental programs implemented;	introduced on the market with a reduction in toxic substances; number of products that can be reused; number of products with a green label;	number of species per unit area;
Indicators for the management area	Indicators for the operational area	Indicators for the environmental area
	Waste : quantity of waste per year or per quantity of product; amount of recyclable or reusable waste per year;	Fauna: concentration of contaminants in animals; number of species identified by area units;
	Emissions : amount of emissions per year; quantity of emissions per unit of product; amount of material dumped in landfills per year;	Man, habitation: human health; erosion caused by constructions;

Source: CAVALIERI (1997).

METCALF et alli (1996, 1996a and 1995), present a series of three papers on the evaluation of environmental performance as a good indicator for benchmarking business progress. The series begins with a "benchmarking[4] ", which identified the best practices, and is characterized by measures as an arrangement of indicators, each indicator having its own units of meaning.

The aforementioned authors also report that the data is collected and put into reports,

4 Technique for recognizing market best practices (Kaplan and Norton, 1997).

representing a certain period of time, presenting a structured report. Each indicator has quantitative targets. Each environmental performance index (IPA) is analyzed through a series of data, and continuous measures are then taken in each IPA area.

On the other hand, some benefits were also observed, such as the following:

• They promote a safe path between the corporation's goals, creating common objectives, giving employees information about strategic directions, or rather where they want to go;

• It defines and supports the corporation's environmental policy, which is the concept of how it considers environmental issues and translates its image. Environmental managers are trained to make decisions in accordance with this policy;

• It highlights areas of excellence in the corporation, encouraging managers to focus on areas of excellence, i.e. areas of improvement for decision-making, always demonstrating continuous improvement;

• Identifies weak points in the environmental management system;

• Provides efficiency in the distribution of resources;

• It provides a communication tool focused on stakeholders.[5]"It can also be a valuable tool for communicating with other business units within the corporation, bringing information, data, vocabulary and real aspects to their attention;

• It provides a mechanism for re-establishing responsibility for results, so that your employees become more responsible and therefore more productive.

Thus, Metcalf (1996a), observing the best environmental practices in companies, incorporated a framework of indicators based on four critical areas:

Environmental Compliance Management - takes into account laws, regulations and codes of business commitment, and your (company's) relationship with government bodies (municipal, state and federal) and your employees;

Training - well-trained employees are the first to take action against environmental problems, eliminating them can be considered a crucial area for the development of the environmental management system;

---

5 They are divided into three categories: those with a financial interest, those with an environmental interest and those with an interest in public policy issues. (Class notes, subject "Strategic planning and the environmental variable", PPGEP-UFSC.

Reduction Programs - is mentioned here in order to capture all the positive environmental activities of the corporation that do not fall under regulation and involve the well-being of the company. Although its benefits are not measured, it is included because of its importance in management;

Effective Environmental Programs - is the analysis of the use of by-products and waste disposal methods; it is also not regulated and its responsibility falls on several people;

ABREU (2001) in his study involving the ECP model (structure - conduct - performance) says that the basic principle of a company's performance lies in the reflection of its competitive practices, or patterns of conduct, which are dependent on the structure of the market in which it operates.

The use of the model proposed by ABREU (op. cit.) determined three types of environmental conduct: the strong, which advocates the existence of a quantitative evaluation of the company's environmental performance; the intermediate, which are companies at the beginning of the implementation of the

quantification of their performance. On the other hand, weak conduct belongs to companies that don't measure, but only do qualitative analysis.

2.4 Mining and environmental impact

The wide variety of environmental problems caused by mining can be categorized into a small number of variables, some types depending on the degree of severity of the occurrence; in this way DOWN & STOCKS (1978) divide them into the following table:

Table 4: Categories of environmental problems caused by mining.

Direct risk to human safety: mining projects present a great potential for dangers and risks to workers' safety.
Indirect risk to human well-being: These are long-term terms and relate to exposure to pollution and noise, which become problems for public health.
Danger to property, agriculture and livestock: These are impacts that are reflected in the
community around the mined area, affecting the quality of production. Discomfort, inconvenience: alteration to human health or safety. Source: DOWN & STOCKS, 1978.

The direct risks to human safety are located in operations, transportation and storage; on the other hand, public safety, as another aspect of the risks, has as its main characteristics: reservoir failures, leaks in waste dams, decantation ponds, landslides on the walls of mines with repercussions in the surroundings of the mined area, especially in abandoned places; leaks of toxic effluents; use of explosives and finally, the transportation of ore.

Indirect risks to human well-being are measures normally adopted to eliminate or minimize the effects generated by mining and which pose an indirect danger to humans, such as air and water pollution by metals and toxic substances, noise and dust, which cause problems for public health. Dangers to property, agriculture and livestock are problems that range from a loss of quality of life to a drop in productivity.

These aspects can be seen in air pollution (emissions of NO_x, SO_x and other gases) and water pollution, which affects the aquatic food chain and causes irreparable damage to the ecosystem; Likewise, subsidence also affects the surrounding area; in addition to the vibrations caused by the explosions and their transmission through the air by the detonation operations of the explosives, causing structural weaknesses and generating rock displacements. Discomfort and inconvenience are difficult criteria to quantify and/or generally accept, and can be classified as visual intrusion, noise and vibration, low levels of air and water pollution and, finally, abandonment of areas to the weather, limiting their potential for use.

SILVA (1988) used the size of the operation as a criterion for answering the question of the extent of the impacts, basing his reasoning on Act No. 045/97 of the Regional Council of Engineering and Architecture - CREA, which sets annual production limits as a delimitation of size, as follows :

Table 5: Annual production limits according to mineral commodity - sand.

Mineral	Small	Average	Large
Sand	≤ 15.000 m 3	15,000 to 40,000 m 3	>40.000 m 3

Source : Act n.045/97, CREA/PR, on 16/09/97.

The sand extractors we want to analyze are concentrated in the first band, i.e. they are small, producing up to 15,000 m 3 per year.

The mining method is also described by SILVA, op.cit., and classified as riverbed suction extraction, which is corroborated by KOPEZINSKI (2000); MINEROPAR (1991) and BRUSCHI and PEIXOTO (1997).

The mineral asset to be exploited, in turn, will determine the entire treatment process, the generation of waste and its by-products, its storage or processing area and disposal operations (SILVA, ibidem).

2.5 The environmental impacts of sand extraction

According to the draft standard of the Brazilian Association of Technical Standards - ABNT (1:602.02-006) of October 1988, which seeks to characterize the polluting loads in mining, the removal of nature causes various alterations in its various phases, as also reported by SILVA (1996), among others already cited.

These phases are mineral exploration, mining (with implementation, operations and construction, as well as the suspension of activities), beneficiation (with implementation and operation) and handling of the sand in the deposit, as well as support activities.

The extractions analyzed are carried out in watercourse beds and, according to BRUSCHI and PEIXOTO (op. cit).These include changes to the landscape, suppression of vegetation, changes to the structure of the soil (compaction, exposure to the sun and changes to the microbiological structure), interference with fauna, conflicts over land use and the use of natural resources, changes to the channels of watercourses, alterations to the water table, shaking, noise pollution, atmospheric pollution, contamination by oils and greases, instability of banks and slopes, water turbidity, liquid effluents, solid waste and alterations to local and regional traffic. This data is confirmed by the environmental impact reports and reports on the recovery of degraded areas, by MACHADO (1998 and 1998 a and 2000); SILVA (1992 and 2000); SILVA and MACHADO (1992).

Corroborating this idea, SILVA and RADOS (2000) drew up a map of the sand extraction process, with a view to understanding the environmental changes that this process could cause (figure 1).

It can be seen that the outputs of the process conceived in figure 1 coincide well with what has already been described by the reports referenced in this chapter, serving as a basis for the proposal to develop a set of indicators for environmental assessment, as well as their limits of application.

Figure 1: Description of the process of extracting sand by dredging in a riverbed.

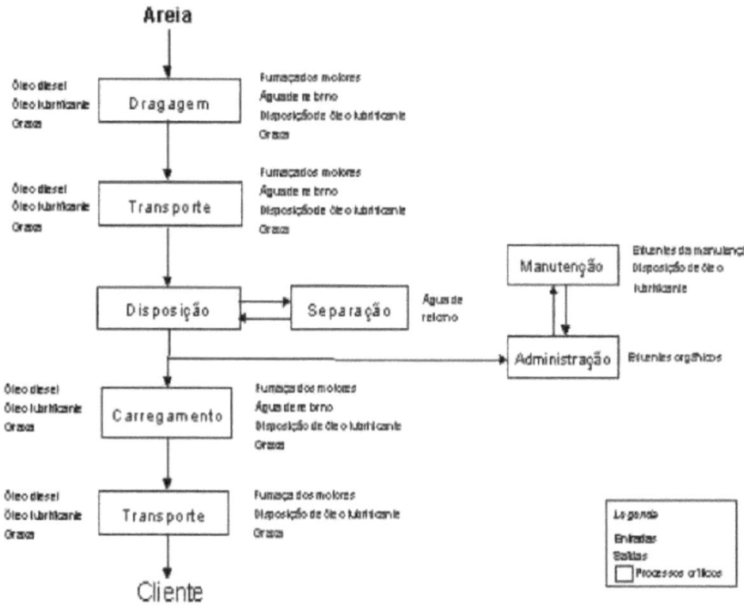

Source: Silva and Rados, 2000

2.6 The set of indicators

CAVALIERI, (1997), recommends that the first aspect to be observed is the "organizational profile", to this discussion can be added an approximation of the size of the operation, whether it operates with other extraction plants in the region or not, whether this is its main line of action and ANDRADE, TACHIZAWA and CARVALHO (2000) puts the nature of the environmental impacts of production. The indices will seek to reflect the trend expressed in performance evaluation methods, especially strategic planning by KAPLAN & NORTON (1997) and more current discussions developed by CARVALHO (2000) and CZAJA (2001).

The latter authors consider that modern management should be proactive, or rather, should consider systematic monitoring in its market action strategies, as well as *benchmarking* with other companies, thus increasing its business competitiveness.

Thus, the trend among large companies is that a complete Environmental Management System must incorporate elements of safety, health and quality, CARVALHO (2000).

CAVALIERI, op. cit., presents a summary of three environmental reporting structures (according to BUHR, PERI and BARTOLOMEU), all three of which agree that indicators should be present in these areas:

Table 6: Set of environmental indicators.

1. Organizational profile: size of the organization, number of plants, branch or main activity and nature of the environmental impacts of production and distribution.

Continuation of Table 6.

2. Environmental policy and management: year of introduction and content, degree of top management involvement and commitment to environmental issues, expectations of achievement, monitoring, level of implementation of the environmental management system, description of training programs, total quality management and continuous improvement.
3. Risk management and legal compliance: identification of risks, description of prevention programs, prevention efforts, description of emergency response programs and how they are communicated to neighboring communities and interested organizations, descriptions of formal risk reduction policies or management practices, description of major accidents, scope and magnitude of environmental impact and programs implemented.
4. Environmental auditing: audit scope and results.
5. Waste: description of waste management, reduction, recycling, treatment, incineration and disposal initiatives (landfill or soil).
6. Atmospheric emissions: legal requirements (in general and on the site) compliance, reduction efforts, quantities, identification of the main responsibilities, emissions affecting the ozone layer and the release of greenhouse gases.
7. Liquid effluents: requirements (in general and by site), compliance, toxicity, effluents by site and identification of responsibilities.
8. Noise and odors: efforts to reduce, exceptions and number of complaints.
9. Product management: new techniques, process and product design, programs for suppliers to reduce impacts and describe the life cycle and its impacts and mitigation measures.
10. Environmental expenditure: definition, capital expenditure (current and comparative), operating expenditure (current and comparative), explanation of accounting criteria and

miscellaneous costs.

Continuation of Table 6.

11. Resource conservation: affected areas, protected areas, material conservation, energy and water conservation, forests and soils, other programs.
12. Employee recognition: information on participation in programs.
13. *Stakeholder* involvement: description of efforts to involve other institutions in environmental initiatives and relations with the public administration, environmental organizations, universities, among others.
14. External certification .
15. External verification of the report.

Source: CAVALIERI, op. cit.

From the above, we can deduce a set of indicators for analyzing the environmental performance of a sand extraction plant using the riverbed dredging method.

Environmental policy: Scope, objectives, targets, date of introduction, validity or revision, or explicit reference to the statement that the company cares about the environment. This aspect aims to analyze whether top management is already concerned with improving its processes, understanding its business, clearly demonstrating the company's mission and explicitly referring to environmental improvement.

It can be seen as how the company considers the relationship between environmental issues and its image, METCALF et all (1996).

Environmental management system: Level of organizational responsibility for environmental policies and management structure (objectives and targets pursued, initiatives implemented and the number of training courses and employees trained).

It is the organizational consequence of the concerns of a clear environmental policy, as it makes it possible to ascertain whether there are people responsible for each task, for each action within the organization, in this way there is a manager responsible for the environmental area, there are people with clear responsibilities for the area of monitoring and measuring the environmental aspects considered relevant by the environmental impact analysis. As CZAJA (2001) observed.

For JOHR (1994) it is imperative to take the lead in the process, as no one becomes "green" overnight, proposing a system that encompasses several aspects.

There are seven factors that must be taken into account (economy, framework, efficiency, education, ethics, engagement and excellence), because taking environmental demands into account will require the business community to do much more than have a cosmetic policy, but to understand how much these issues involve their business and can also contribute to their profits.

To manage a complex chain of interrelated facts, from the manufacture of the product to its suppliers, customers, employees, the media and the community, obtaining synergy in the results (CHEHEBE, 1998). In this way, to analyze whether there is regular training with a concern for quality, worker safety and environmental quality that focuses mainly on continuous improvement; in addition, to observe whether there are regular environmental audits in the company.

Risk management system: Description of accidents and risks, description of prevention programs and discussion of the most significant activities, such as mapping the use of hazardous waste (quantity and cost of disposal), number of chemical deposits, introduction of alternative fuels (biomass burning, recycled oil, other types of fuel).

The risk management system includes not only the requirements of Law No. 6.514/77, the Environmental Risk Prevention Plan - PPRA, but also those related to goods and services purchased, so that operational controls can be extended to suppliers with the development of the management system, whether in recycling, the use of alternative fuels or many others.

In addition, for altered environmental aspects and occupational risks involving emergency situations, there are provisions for preparing and taking measures to deal with emergencies and even mention of critical analysis of the performance of these systems after testing or emergency occurrences, as well as regular audits, according to De CICCO (1995), CARVALHO FILHO (1998), GREGORIS (1994), FUNDACENTRO (1981) and SILVA (2000a).

GALVÀO FILHO (2001) states that these studies are defined as a process of estimating the probability and occurrence of an event and the probable magnitude of its adverse effects on health, human safety and the environment, over a specific period of time.

So analyze whether this report (PPRA) exists and its development, such as courses, training, and critical analysis with regular audits.

Legal compliance: Number of lawsuits, number of fines, compliance reports and nature of non-conformities, belonging to the first level of KINLAW, op.cit., and should also include integrated plans and initiatives and a series for data analysis METCALF and others, (1995, 1996 and 1996a), such as number of notifications from the environmental agency (prefecture, state or federal) or number of violations and total in currency, in fines.

It concerns the entrepreneur who implements a legal management system for his business. Does this entrepreneur realize exactly how much money is spent on legal proceedings, fines and notifications from the environmental agency, METCALF et all (1996).

Product management: Product innovations (eco-efficiency, environmental design, among other names, CALLEMBACH, op. cit.), cooperation program with customers and consumers (number of programs implemented and number of operational strategies in environmental quality, METCALF et al.), number of people from the community involved, *stakeholder* involvement, critical evaluation of programs), discussions on new techniques and design for products and processes, waste reduction (gaseous emissions such as SO_2, NO_x, CO_2, among others) being considered as positive activities outside of regulation; material recycling efforts (amount of metal, paper and cardboard, wood and the cost associated with the benefit), energy use efficiency efforts.

It seeks to analyze the production and support processes that are essential for making the product with a view to environmental design for the production line, with a focus on customer satisfaction (internal and external), involving design and development (controls, validations, identification and traceability), the acquisition of new technologies and product preservation. The basic idea is to incorporate the ideas of eco-efficiency, which encompass economic and ecological resources, with productivity (doing more with less), waste reduction and, concomitantly, less pollution, into the entrepreneur's day-to-day life, creating additional value to meet their needs, according to SIMONE (1997).

Other programs: This section seeks to increase the effectiveness of the information provided above, complementing what we have seen with the programs implemented by the company's environmental participation. These are reflections of the ideas of the entrepreneurs, the policies and strategies issued and implemented by the board of directors.

These can be presented as environmental education programs carried out or financed, environmental competitions, awards and competitions that are also ecological in nature; the number of participants in government or NGO (non-governmental organization) environmental projects in the mining areas or their surroundings. Financial performance: relates to everything that can be compared financially, or rather, has a monetary relationship with environmental or management variables:

$[6] total in adjustment projects/$ total in fines, (measured in the fiscal year) is an index that seeks to demonstrate the growth in investments over the decrease in fines.

6 $ - Refers to the quantity in currency.

42 $ total investment in environmental programs/$ total investments in the period, (measured in the fiscal year) is an index that seeks to demonstrate the increase in environmental investments over the company's total investments.

$ in electricity spent/ton of sand (can be measured per month or per fiscal year) and seeks to show the decrease or increase in energy expenditure per ton of sand that the mine produces.

$ in oil-based fuel spent/tonne of sand (can be measured per month or per fiscal year) shows the decrease or increase in the use of fuels from the oil matrix per tonne of sand produced.

$ fleet consumption/tonne of sand (can be measured per month or per fiscal year) shows the increase or decrease in fuel consumption.

It is a block that seeks to analyze the relationships in currency with turnover and investments, in percentage, and spending on environmental projects and programs. KONAR and COHEN (2000) cite that their studies relate a company's financial performance to its environmental concern, since poor accounting reveals negative effects on intangible values, such as brand, credibility, among other aspects.

In another study, COHEN, FENN and KONAR (1997) point out that although environmental regulation is costly, it leads to efficient control of production and that they compare the market return between small companies, dividing them into highly polluting and low-polluting, where the latter have more investors ready to put their money into these businesses.

On the other hand, how much has been spent on resources (electricity, fuel, etc.) and their relationship to the quantity produced is information that will help make an environmental control program a success, avoiding excessive spending that could harm the company.

Accounting is still in the process of discussing an information system with which to deal with environmental issues. MARTINS and De LUCA (1993) define environmental assets as all the company's assets aimed at preserving, protecting and recovering the environment (i.e. stock, fixed assets, deferred charges, provisions for devaluation and accelerated depreciation), while CARVALHO, MATOS and MORAES (eds.) define environmental liabilities as all known and measurable obligations due to the expectation of future spending, taxes and regulations, as well as all environmental liabilities.), conceptualize environmental liabilities as all known and measurable obligations due to the expectation of future expenses, taxes, environmental legislation and regulations, as well as all obligations arising from environmental issues.

With these concepts, FERREIA (2002) elaborates an intricate accounting information system and proposes the environmental issue as a source of management concern, providing a model that describes operational and financial economic events.

Finally, from the accounting point of view, NORTH (1992) and HOPFENBACK (1993), apud CARVALHO (2000) introduce some environmental control criteria for the company:

Units produced/electricity consumed;

Units produced/waste produced e;

Productive losses caused by environmental problems/period of time.

CARVALHO, (2000), proposes the use of eco-efficiency indicators, more based on costs and health:

Environmental cost of production - CAP

It measures the relationship between environmental costs and the total cost of production, by separating the costs of preserving and repairing damage to the environment from the total costs of production over a period of time; it defines a percentage of the expenses incurred by the company and its concern for sustainable development.

Environmental cost unit - UCA

The relationship between the environmental cost of production and the quantity of products manufactured in a given period can be evaluated using the formula:

UCA = CAP/UPP

Where :

UCA - Production cost unit;

CAP - Environmental cost of production;

UPP - Units produced in the period.

The aim is to determine the environmental cost per unit produced, with the possibility of comparing it with the incremental revenue provided by the value of "green products" and then comparing the additional revenue with the cost and additional revenue resulting from the adoption of measures to conserve and preserve natural resources.

Natural energy consumed - ENC.

It is the great challenge of accounting to measure renewable and non-renewable resources. It measures the ratio between the consumption of renewable and non-renewable natural resources in relation to the quantity of products manufactured in a given period:

ENC = RNR+RNNR/UPP

Where:

ENC - natural energy consumed;

RNR - renewable natural resources;RNNR - non-renewable natural resources;

UPP - units produced in the period.

Environmental development or, as CAVALIERI (op.cit.) puts it, environmental compliance management: produces a database of environmental data relating to sand extraction, its process and its surroundings.

It is based on the requirements found in the environmental impact reports of SILVA (2000) and MACHADO (2000), the environmental recovery plans in SILVA (1992), SILVA and MACHADO (1992), SILVA (1996) and MACHADO (1998 and 1998a) among other citations, such as KOPEZINSKI, (2000), ABNT (1996) and BRUSCHI and PEIXOTO (1997).

Its limits consist of legal parameters set out in federal, state and municipal laws, as well as federal resolutions (CONAMA, e.g.), which are listed below for a better understanding:

Air (contaminants, noise, odors);

Water: (concentration of chemical and biological contaminants, BOD 5 days, DO (dissolved oxygen), dissolved coliforms/ liter of water);

Soil: erosion (hotspots or areas prone to erosion);

Flora: vegetation (quality and quantity);

Species (quantity/unit area, e.g. quantity/m^2);

Fauna: (quality and quantity), number of species/area unit.

Men (in the sense of people): health indicators (at the extraction site, with the employees and in the area surrounding the extraction site), such as vaccinations, occurrences of diseases linked to public health, among others;

Housing indicators (quality, water and sewage services, urban criteria such as asphalt, health centers, schools, among others).

In this proposal, or rather, for each process diagnosed, there is a geographical boundary that may or may not be based on the location of the enterprise, such as a point on the map, a point in the production process, or even a certain area that represents a geographical situation CHEHEBE (1998), In this way, these limits can be established by the area requested as a mining area from the National Department of Mineral Production - DNPM (an agency of the Ministry of Mines and Energy for the administration of the Brazilian subsoil) or even be considered another extension for its analysis.

Its limits consist of legal parameters set out in federal and state laws and federal resolutions (CONAMA), as well as various authors who describe environmental assessment methods, which are shown below:

Air :

CONAMA Resolution No. 10, of 14/09/1989, which sets limits for exhaust emissions from diesel vehicles;

CONAMA Resolution No. 03, of June 28, 1980, establishes standards for air quality;

CONAMA Resolution No. 08, which sets maximum emission limits for air pollutants from fixed pollution sources.

Noises :

CONAMA Resolution No. 01 of March 8, 1990, which establishes compliance with noise levels for acoustic comfort (NBR 10.151) and by the Ministry of Labor NR (Law No. 6514 of December 22, 1977, Regulatory Standards).

CONAMA Resolution No. 01, of 11/02/93, establishes maximum noise limits for vehicles under acceleration and at a standstill.

Odors

Law No. 6514, of December 22, 1977, regulatory standards.

Water

Contaminants :

CONAMA Resolution No. 20, which sets limits for water quality.

Soils

Erosion hotspots: present or not, according to MEDEIROS, (1995:7); GUERRA and CUNHA (1996).

Suitable areas: Geotechnical geological mapping, using the method described in MEDEIROS, (1995:5).

Flora

Quality: Description of vegetation present.

Quantity : Mapping according to MEDEIROS, (1995: 26).

Fauna

Quality: description of avifauna, herpetofauna and ichthyofauna; vector invertebrates and mastofauna.

Quantity : mapping according to MEDEIROS (1995)

Man

Health and Vaccination: Survey of secondary data and observation.

Diseases related to public health: related to contamination, water and sewage, vectors, health at work. MEDEIROS, (1995:58).

Housing

Number of units, occupancy conditions and type of construction (according to MEDEIROS,

1995:51).

2.7 General considerations

From these observations, it appears that small mining companies, when considering a modern form of management, must take into account the evaluation of environmental performance as a critical success factor.

In this way, environmental performance fulfills the function of not only describing the natural environment, but also promoting integration between tangible and intangible measures, providing a new image for the sand mining company, which goes from being a villain to a friend of the environmental cause.

It also acts to transform the entrepreneur and his or her management, encouraging a wide-ranging discussion of the strategy to be adopted in business, in corporate life and in relations with employees.

Chapter 3 - A proposal for environmental performance indicators

3.1 Environmental performance indicators

On the previous pages you can see that one of the best ways to analyze environmental performance is by producing a series of indicators.

The company's environmental management already occupies a prominent position in organizational functions, not only because it contributes positively to the image of quality, but also because of its relationship with improving the production process, reducing costs and being part of the total quality policy, VALLE (1995).

Environmental management must be committed to expressing its intentions and principles in its environmental performance report.

It must therefore be proactive in its market action strategies, according to CZAJA (2001), especially considering systematic monitoring, benchmarking and business competitiveness, according to CAVALIERI (1997) and CARVALHO (2000). This attitude is manifested in a system that incorporates elements of safety, health, quality and natural environment issues.

With this in mind, the GERMAN ENVIRONMENTAL MINISTRY (1997) has developed three main groups of environmental indicators, which, when adapted to the current study situation, are as follows:

- Management indicators;
- Financial indicators and
- Indicators of environmental conditions.

Management indicators show the organization's actions to minimize negative environmental impact; they serve as an internal control, although they do not reflect the impacts caused by the company and cannot be used alone to measure environmental performance.

These are: environmental policy, the environmental management system, the risk management system, the legal management system, product management and environmental education programs.

Financial indicators are elements of the organization that seek to define and relate financial variables to environmental variables; they want to analyze the relationship in terms of currency, turnover and investments, as well as spending on environmental programs and projects. These are: total fines, total investments, total spending on electricity and its relationship to production, total fuel spent and its relationship to production. There are currently two contexts for analysis, when the company applies the social balance sheet, SILVA (2001) and when it doesn't use this instrument, this being the most common situation.

Environmental condition indicators seek to demonstrate the relationship between the sand extraction process and its surroundings through the creation of a database; they are the negative impacts of extraction. They are based on environmental impact reports and environmental control plans.

All these indicators are important, because they are systemically related to each other, and one does not take precedence over the other.

These indicators are :

Air (emissions, noise and odors);

Water (contaminants and their concentration, DO - dissolved oxygen, coliforms present);

Soil (erosion);

Flora (diversity);

Fauna (diversity);

Men: health indicators; housing quality indicators.

3.2 Overview of the proposed model.

This model proposes indicators for evaluating the environmental performance of a sand port using the dredging method on a riverbed.

In addition to the three main groups of indicators (management, financial and environmental conditions), this work also provides for their characterization as qualitative, according to CZAJA (2002), when they appear in this model in three states: they are present, they are not present or they are in the process of being implemented.

On the other hand, there are indicators that determine levels, or relative quantities, still according to CZAJA (op. cit.), establishing limits, which can be called quantitative.

These environmental performance assessment indicators (both qualitative and quantitative) can be highlighted as :

- Environmental policy (qualitative);
- Environmental management system - EMS (qualitative);
- Risk management system - SGR (qualitative);
- Legal management system - LMS (qualitative);
- Product management (qualitative);
- Environmental education programs (qualitative);
- Financial indicators (quantitative);
- Indicators of environmental conditions (quantitative).

Each of the above indicators is highlighted below.

3.2.1 Qualitative management indicators

These indicators are :

- Environmental policy;
- EMS;
- SGR;
- SGL;
- Product management;
- Environmental education projects.

Using a deterministic classification for understanding and weighting.

3.2.1.1 Environmental policy

CAVALIERI, (op. cit.) reports, through the study of three environmental performance reporting structures (BUHR, PERI and BARTOLOMEU), that all agree on presenting environmental policy as the initial indicator.

This indicator reflects the commitment of management (owner, partners and/or shareholders) to establishing a set of clear and defined guidelines that will guide the organization's environmental activities, VALLE (1995).

In addition, they provide a working context for setting the organization's goals and objectives, TIBOR and FELDMAN (1996).

The policy can take into account the organization's mission, core values, stakeholder requirements and specific local and regional conditions, while responding to the financial requirements of the business.

MOURA (2002), TIBOR and FELDMAN (1996) and VALLE (1995), among many other authors, draw up an environmental policy based on some basic premises, which are :

- that it is appropriate to the nature, scale and environmental impacts of the organization's activities, products and services;
- which includes commitments to continuous improvement;
- which includes a commitment to pollution prevention;
- which includes a commitment to comply with legislation, regulations and other relevant requirements to which the organization is subject;
- that provides a contextual working framework for setting and reassessing environmental objectives and targets;
- that it is documented, implemented, maintained and communicated to all employees;
- to educate and train its employees so that they always act in an environmentally correct manner;
- and that it is available to the public.

The aim of this dissertation is to evaluate its scope, objectives, goals, date of introduction, validity and revisions. METCALF et all (1996) put environmental policy as a way of relating environmental issues to the company's image.

In this work, the variable associated with environmental policy will be treated as a deterministic variable and the values can be between three situations:

- environmental policy exists - YES;
- environmental policy doesn't exist - NO;
- the environmental policy is being implemented - IN IMPLEMENTATION.

3.2.1.2 Environmental management system - EMS.

For JOHR (1994), the environmental management system should propose a system that encompasses several aspects, taking into account seven main factors (economy, framework, efficiency, education, ethics, engagement and excellence), requiring much more than a cosmetic policy. This indicator aims to show the consequences of the environmental policy's positions, allowing us to gauge whether there are people responsible for each task, for each action within the organization.

This means checking whether there is a manager responsible for the environmental area, whether there are people with clear responsibilities for monitoring and measuring the environmental aspects considered relevant by the environmental impact analysis, CZAJA (2001).

The environmental management system, in addition to the establishment of an environmental policy, presupposes action planning, a training and motivation strategy in relation to its environmental performance, the establishment of procedures and work instructions (environmental management manual), which results in ongoing work aimed at continuous improvement, in a structured process of verifications and monitoring of all the environmental variables considered important in the process of extracting sand from riverbeds using the dredging method.

In this way, CHEHEBE (1998) sees EMS as the management of a complex chain of interrelated facts, from obtaining the product to its suppliers, customers, employees, the media and the community, thus achieving synergy in the results.

It also looks at whether there is regular training for employees in the area of environmental quality, worker safety and health; finally, it analyzes whether all these activities are regularly audited by the company.

As with the previous indicator, this variable is associated with the same three stages, namely :

- the EMS exists - YES;
- the EMS does not exist - NO;-

the EMS is being implemented - IN IMPLEMENTATION.

3.2.1.3 Risk management system - SGR

This system promotes the control and minimization of pollution sources and the correct disposal of waste generated by the company, incorporating two components, the probability of occurrence and the severity of potential damage.

The level of a risk can be assessed according to the frequency with which risk situations occur and the severity of the resulting effects. It is an important tool for identifying vulnerable points in a facility, GALVÀO FILHO (2001) or a process, allowing preventive measures to be adopted to protect the environment and humans.

The SGR provides not only for the requirements that are part of law no. 6.514/77 - the environmental risk prevention plan - PPRA, but also for those related to goods and services purchased, FUNDACENTRO (1981), where its operational controls extend to suppliers, whether in recycling or in the use of alternative fuels.

In addition, for altered environmental aspects and occupational risks that involve emergency situations, there will be provisions for preparation and training to deal with these emergencies, DeCICCO (1995), CARVALHO FILHO (1998) and GREGORIS (1994).

These are systems that describe the accidents that have occurred and the associated risks, guided by prevention programs, mapping of the most significant activities, such as waste production (quantity, cost of disposal, among others), chemical product deposits, discussions about the introduction of alternative fuels (biomass burning, recycled oil and other types of fuel).This variable, also deterministic, is associated with three situations, which like the previous ones are :

- the SGR exists - YES;
- the SGR does not exist - NO;
- the RMS is being implemented - IN IMPLEMENTATION.

3.2.1.4 Legal management system - LMS

It concerns a company that implements and maintains a legal management system for its business. METCALF et all (1996) point out the size of their relevant expenditure on legal proceedings (one of the elements of their liabilities), fines and notifications from government bodies.

This indicator also includes a training system for assessing legal compliance and understanding case law on the subject.

Among the most relevant data are :

- the number of court cases;
- the number of fines, notifications from the environmental agency, violations;
- presents the conformity report, describing the nature of the non-conformities;
- presents integrated plans and initiatives for achieving legal compliance.

Like the previous ones, this indicator is also deterministically framed in :

- there is LAS - YES;
- there is no LAS - NO;
- the LMS is currently being implemented.

3.2.1.5 Product management

It seeks to analyze the production and support processes that are essential to the realization of the product, with a view to environmental design, with a focus on customer satisfaction (internal and external). It also involves design and development (controls, validations, identification and traceability), the acquisition of new technologies and product preservation.

The basic idea is to incorporate the ideas of eco-efficiency into the day-to-day life of the company and the entrepreneur, encompassing economic and ecological resources, with productivity (doing more with less), reducing waste emissions, generating less pollution and thereby creating additional value for the product and its needs, according to SIMONE (1997).

As CALLEMBACH (op. cit.) tells us, product management seeks to demonstrate product innovations (eco-efficiency, environmental design, among other names), cooperation programs with customers and consumers, typification of the programs implemented and the number of operational strategies in environmental quality, METCALF et all (op. cit.) also cites the number of people from the community involved in projects, stakeholder involvement and a critical evaluation of the programs.

Product stewardship also features discussions on new techniques and new designs for products and processes, as well as the reduction of waste (gaseous emissions such as SO2 , NOx , CO2 , among others).

Positive activities, outside of regulation, are efforts to recycle material (amount of metal, paper and cardboard, wood and the cost associated with the benefit), as well as efficiency in the proper use of electricity.

And, like the other indicators already mentioned, they are classified as :

- there is a concern for product management in place - YES;
- there is no concern for product management in place - NO;
- this is placed in such a way that it is considered to be in implementation - IN IMPLEMENTATION.

3.2.1.6 Environmental education programs

These are considered to be educational development programs involving the environmental component that the company can present, such as its participation in community events.

It is also a reflection of the company's ideas and environmental policy, going beyond its strategies, being an analysis that describes the environmental education programs that have been, or are being, carried out and/or financed by the environmental company.

They can be related to awards or competitions with an ecological meaning. If they are being held in partnership with the government or NGOs, if they involve the community or if they take place in areas around the extraction site or in a larger area.

And also, like the previous ones, placed as :

- there are educational development programs in place - YES;
- there are no educational development programs in place - NO;
- is thinking of starting a project along these lines - IN IMPLEMENTATION.

3.3 Quantitative financial indicators

They are financial data that can be correlated, so they are looking for relationships in terms of currency (turnover and investments) and spending on environmental projects and programs. In their studies, KONAR and COHEN (2000) cite the relationship between a company's financial performance and its environmental concern, pointing out that incorrect accounting work has negative effects on tangible values such as brand, credibility and so on. In another text, COHEN, FENN and KONAR (1997) draw attention to the fact that although environmental regulation is costly, it leads to efficient control of production.

MARTINS and De LUCA (1993) state that environmental assets are all the company's assets aimed at preserving, protecting and recovering the environment (fixed stock, deferred charges, provisions for devaluation and accelerated depreciation).

In this way, CARVALHO, MATOS and MORAES (n.d.) conceptualize environmental liabilities as all obligations known and measurable by the expectation of future expenses, taxes, environmental legislation and regulations, as well as all obligations arising from environmental issues. With this discussion in mind, FERREIA (2002) elaborates an intricate system of accounting information, proposing the environmental issue as a source of management concern, providing a model that describes economic, operational and financial events, inserted in a modern accounting context, which is the social balance sheet, KROETZ (2002) and SILVA (2001).

Within the scope of the social balance sheet, NORTH (1992) and HOPFENBACK (1993), apud CARVALHO (2000), introduce some environmental control criteria, which are :

- units produced / energy consumed;
- units produced / waste generated e;
- production losses caused by environmental problems / period of time.

CARVALHO (2000), on the other hand, proposes the use of other eco-efficiency indicators, which are more cost-based and can be taken in the context of the social balance sheet:

<u>Environmental cost of production</u> - CAP : measures the relationship between environmental costs and the total cost of production, defining a percentage of the expenses incurred by the company and its concern for sustainable development. It separates the costs of reserving and repairing damage to the environment from total production costs over a period of time.

<u>Environmental Cost Unit</u> - ECU: is the ratio between the environmental cost of production and the quantity of products manufactured in a given period. It is evaluated using the formula :

UCA = CAP / UPP, where:

UCA - environmental cost units;

CAP - environmental cost of production;

UPP - unit produced in the period.

This formula calculates the environmental cost per unit produced, which makes it possible to compare it with the incremental revenue provided by the proposed value of "green products".

Natural energy consumed - NCE: this is the great challenge of the social balance sheet and accounting, to measure the relationship between the consumption of renewable and non-renewable natural resources, in relation to the quantity of products manufactured in a given period, given by the formula :

ENC = (RNR + RNNR) / UPP, where :

ENC - natural energy consumed;

RNR - renewable natural resources;

NNR - non-renewable natural resources;

UPP - units produced in the period.

The indicators presented above have their own meanings in the context of a social balance sheet. In the case under study, they could be analyzed from another, less modern, accounting perspective, so that they can represent what needs to be described about the "ports of sand".

As such, the following indicators are also part of the proposed model and have been developed to enable a measurement that is closer to the reality of the sand extraction process and its environmental impacts, in order to enable efficient management:

Ratio of adjustment projects to fines - this is the total amount of money invested in adjustment projects divided by the total amount spent on fines in the fiscal year. It is an indicator that seeks to demonstrate the growth of investments over the decrease in fines;

Investment ratio - Total amount invested in environmental programs / total amount invested in the period, in the fiscal year - is an indicator that reflects the increase in environmental investments over the company's total investments;

Ratio between electricity costs and production - Total amount of electricity costs / amount of sand produced, in tons - can be measured per month or fiscal year, seeks to demonstrate the decrease or increase in energy costs, per ton of sand that the extraction produces;

Ratio between fuel expenditure and production - Amount in currency spent on oil-based fuel / ton of sand produced - as in the previous case, shows the decrease or increase in the use of fuel from the oil matrix, per ton of sand produced;

Ratio of fleet costs to production - Total fleet costs/tonne of sand produced - measured in the same way as above, it shows the increase or decrease in fuel consumption and maintenance by the fleet.

It is analyzed in the questionnaire (in Appendix 1) by the list of questions that ask if there are any indicators present in the traditional balance sheet or even outside of it (when applicable in the social balance sheet). It is not stipulated that the questionnaire can be answered in a quantitative way (which would be the best for this research) but if some of these concerns can be observed in the balance sheet, or even outside of it, this should be analyzed

3.4 Indicators of environmental conditions

It is a set of indicators that will produce an environmental database that relates to the process and its surroundings.

They are based on the requirements of the environmental impact reports (EIA-RIMA) and the environmental control plans (PCA), among the various citations that can be found here.

found, we highlight KOPEZINSKI (2000), ABNT (1996), BRUSCHI e

PEIXOTO (1997), SILVA (1996, 1997), SILVA and RADOS (2000).

Its limits consist of legal parameters set out in federal, state and municipal laws, including federal resolutions, such as CONAMA, which are described below:

3.4.1 Air

Three criteria will be used for air: emissions, noise and odors:

<u>Emissions</u> - the emissions standard and its maximum limits are set out in CONAMA Resolution No. 10, of 14/09/1989, (with the diesel cycle) and CONAMA Resolution No. 10, of 14/09/89, which sets emission values for the two cycles (Otto and diesel). For diesel and Otto cycle engines, the indicators proposed for measurement will be carbon monoxide, hydrocarbons and nitrogen oxides.

Noise: this is in accordance with CONAMA Resolution No. 10, of 1409/1989, which establishes compliance with noise levels for acoustic comfort (NBR 10.151) and the Ministry of Labor, in its Regulations (Law No. 6.514, of December 22, 1977).

<u>Standards</u>

Noise levels will be measured in decibels (Dba) and, according to CONAMA Resolution No. 01 of 11/02/1993, in the vicinity of vehicles (determining maximum limits for stationary and accelerated conditions) and in the working environment, observing the provisions of the standard.

Odors: Law no. 6.513 of December 22, 1977 (regulatory standards) only comments on whether or not they exist on company premises. It should be noted that this is a qualitative indicator.

3.4.2 Water

For water, the quality limits are set by CONAMA Resolution No. 20 of 18/06/1986.

To this end, it is referenced with parameters that report its classification (sweet, brackish, saline and special) and its classification (establishing the level of quality to be maintained).

Thus, for waters, the following are required :- special class - fresh water - the absence of coliforms;

Class 1 - Freshwater - floating materials, oils and grease, substances that communicate taste or odor, artificial colorants, coliforms, 5-day BOD, DO, turbidity, color, pH and other potentially harmful substances;

Class 2 - fresh water - the above limits, plus artificial colorants that cannot be removed by a physico-chemical process, coliforms, color, turbidity, 5-day BOD and DO;

Class 3 - Freshwater - floating materials, oils and grease, substances that communicate taste or odor, artificial colorants that cannot be removed, substances that form objectionable deposits, coliforms, 5-day BOD, DO, turbidity, color, pH and other substances;

Class 4 - freshwater - floating materials, odor and aspects, oils and grease, sedimentable substances, phenols, DO and pH;

Class 5 - saline waters - floating materials, oils and grease, odor-producing substances, turbidity and objectionable deposits, artificial dyes, coliforms, 5-day BOD, DO, pH and other harmful substances;

Class 6 - saline waters - floating materials, oils and grease, substances that produce odor, turbidity and that form objectionable deposits and artificial dyes, coliforms, 5-day BOD, DO, pH;

Class 7 - brackish water - BOD 5days, DO, pH, oils and grease, floating materials, substances that produce color, odor and turbidity, and that form objectionable deposits, coliforms and the presence of potentially harmful substances;

Class 8 - brackish water - pH, DO, oil and grease, floating materials, substances that produce color, odor, turbidity and easily sedimentable substances and coliforms.

3.4.3 Soils

With regard to soils, it is possible to observe (qualitatively) the existence or not of erosion hotspots, or rather the presence of furrows and gullies and deposits of sediment transported by the erosive action of water, according to work by MEDEIROS (1995:07) and by GUERRA and CUNHA (1996); in addition, it is also possible to locate suitable areas through geological-geotechnical mapping (cross-referencing information on the slopes of the site with the geology, type of soil, rainfall, among others), also described by MEDEIROS, op. cit. cit;

3.4.4 Flora

As for the terrestrial flora, a description (qualitative analysis) of its current reality is carried out, representing each vegetation type (herbaceous and shrubby strata), followed by its mapping (measures of density, distribution, dominance, value of importance, floristic similarity indices, diversity indices and rare species indices) according to the basic criteria of MEDEIROS (1995:20).

3.4.5 Fauna

Using the same procedure as above, the fauna is described, such as avifauna, herpetofauna and ichthyofauna, invertebrates and mastofauna, in qualitative terms, with a view to drawing up a preliminary list of occurring species.

Quantitative characterization is observed in fieldwork, with the occurrence of endemic, large, small and medium-sized forms, their frequency and

abundance, in arboreal forms or not, the presence of endangered species, the occurrence of venomous species of snakes and the relationship between these species and human presence (synanthropic and alloanthropic), which should be related to the different environments present in the areas to be worked, according to MEDEIROS (op. cit.).

3.4.6 Human conditions

In the analysis of man's conditions, :

Firstly, public health, in terms of vaccinations, where we check whether the mining company has carried out any surveys on this issue; whether it reports any illnesses related to water, noise or atmospheric pollution, or whether it has any notes on water and sewage contamination, or even the presence of disease vectors (rats, cockroaches, among others), or even sexually transmitted diseases, among other aspects of health and safety at work.

As for housing, it can be seen whether there are units and the number of dwellings, their occupancy conditions and the method of construction (wood, blocks, rammed earth, among others), according to MEDEIROS (ibidem), and it should be noted that this is also a qualitative index.

These indices are obtained through a survey supported by the health departments of the municipalities where the extractions are located. In all the indicators presented, there is a geographical limit, which may or may not be based on the location of the sand extraction (as a point on the map) or a site in the production process, or even in a broader geographical situation, CHEHEBE, (1998).

For this work, the observation horizon adopted was the area requested as a "mining area" by the National Department of Mineral Production - DNPM, an agency of the Ministry of Mines and Energy, which manages the Brazilian subsoil.

3.5 Data collection instrument

3.5.1 Introduction

Based on the literature review (previous chapter), we tried to obtain a series of environmental performance indicators for sand extraction using the dredging method in riverbeds. With this in mind, a questionnaire (appendix 1) was drawn up to provide guidance on the indicators to be researched, opting for an open-ended format, although some limits were observed, which tended to facilitate discussion and the obtaining of information.

3.5.2 Some considerations

This study does not take into account the second-order or greater effects produced by the environmental impacts on the mining areas and their surroundings, due to the difficulty of measuring the chain of causes and effects on the environment.

The observations and interpretations of the characteristics of sand mining in riverbeds by dredging are multidisciplinary.

Thus, the presentation of the data should generate a discussion, leading to comparisons and considerations, which will lead to the continuous improvement of its environmental performance.

On the other hand, it is hoped that these indicators can be robust in the analysis of environmental impacts, that they can be complementary, inclusive, in other words, that they can serve to explain each other, forming a whole that will discuss a level of excellence for sand mining in riverbeds by dredging.

Its continued use could enable the company to become a sector benchmark for mining.

Chapter 4 - Validation of the set of indicators for sand extraction by the dredging method in riverbeds.

4.1 Introduction and criteria for validating indicators

Once we had the list of criteria described in the previous chapter and put them into the form of a data collection tool, we went on to apply them to two sand mines in the municipalities bordering the Paranà River, belonging to the Environmental Protection Area - APA, managed by the Intermunicipal Consortium for the Conservation of the Remnants of the Paranà River and Areas of Influence - CORIPA, observing their production processes, ways of managing the business and, above all, their relationship with the community. The original names of these mines have been deliberately omitted and they are referred to only by the municipality to which they belong.

The technicians in charge of the two mining companies were then given a questionnaire (Appendix 1), which allowed them to answer questions about the environmental situation of the mining companies, the quality of the measurements being proposed and the type of actions being taken in the environmental field.

4.2 The data obtained by applying the questionnaire.

The data obtained from the questionnaire at the two mines chosen to validate the ADA proposal is presented below.

4.2.1 Mining in the municipality of Guaira, PR.

The person who received the questionnaire and was the respondent, works as a professional in mining technical advice. It was carried out in the company's office, surrounded by half-answers and a lot of fear. Some concepts had to be explained more than once, even though the questionnaire had been presented beforehand for consideration.

What emerged is that the questionnaire wasn't even read.

The following aspects stand out.

Indicator 1 - Environmental Policy: The company does not have an explicit, written environmental policy that all employees are aware of. It only reacts when pressured by the environmental agency.

Indicator 2 - Environmental Management System - EMS : There is no structured EMS, actions are taken according to political pressure and/or legal pressure. The board of directors is at a stage of reactivity to environmental issues.

Indicator 3 - Risk Management System - SGR : The company has an Environmental Risk Protection Plan - PPRA, recommended by Law 6514, of December 22, 1977, Ordinance 3214, of July 8, 1978 and amendments. This report was drawn up by a safety technician and describes the risks to worker safety, but leaves something to be desired in its implementation.

There is no contingency plan for accidents (fires or explosions, among others), nor studies into the use of alternative fuels, as a plausible environmental solution for the safety of employees, the surrounding environment and the economic health of the company.

Indicator 4- Legal Management System - LMS : There is no organized archive of legal cases. The person responsible had the office of the lawyer who represents the company in legal disputes consulted. When asked, the lawyer said "that he couldn't remember at the time and that he needed to do a survey".

The administration was unable to report on the amount spent on fines in general, and even less on expenses related to environmental issues. It did not present a report on environmental compliance or non-compliance.

There was no information about the notifications from the environmental agency, the labor department and others.

Indicator 5 - Product management: They only know how to extract sand, although the possibility of new products (differentiation in grain size) has been raised in order to achieve a better price for the final product.

There is no other design, according to the board, for the extraction work to be carried out.

There are no regional cooperation projects.

There are no service quality and/or environmental quality maintenance projects involved in the creation of new products.

There are no people from the community involved in mining projects.

There is no critical evaluation, either by the board or by the technical and administrative staff, of possible programs to be implemented. The board only meets for internal financial and administrative decisions.

There are no waste reduction programs, reacting to mass government campaigns such as saving electricity, separating household waste, among others.

Indicator 6 - Environmental education programs: There are no environmental education programs for the internal public, nor training programs in occupational safety.

They are unaware of government environmental programs (federal and state) and only recognize the municipal environmental department and, as a state agent, the Paranà Environmental Institute - IAP.

In relation to the activities carried out by Non-Governmental Environmental Organizations, they still have a very negative image, being seen as enemies rather than helpers in the process of developing the region.

Indicator 7 - Financial indicators : The financial performance measure is a big unknown in this research, where information is denied, saying that it doesn't exist, that it can't be disclosed and that it is only of interest to the company.

We don't know for sure the number of units produced, translated into tons of sand, and the ratio of production losses caused by environmental problems, which means that the indices linked to the CAP, the UCA and, consequently, the ENC are not known.

They also don't show the amount of money spent on environmental adjustment projects, which undermines the index that relates to the fines imposed.

Movement/investments → the answer is - "don't know, better check the annual mining report".

Spending on environmental projects/investments → can't say.

Expenditure on electricity/amount of sand produced → doesn't know, thought it was funny.

Fuel expenditure $/amount of sand produced → can't tell how much sand is produced and need to find out how much diesel is used.

Expenditure on fleet consumption (trucks, boats and equipment) → no exact figure, only approximate.

Quantity of sand produced → the answer is - "it's difficult to know, it's better to look in the annual mining report".

There is no interest in this type of measurement, because they cannot interpret a "value" in this data, and accounting professionals are not prepared to evaluate this data. There are no conditions for measuring waste in order to make the relationships.

Thus, the amount produced, the amount of electricity consumed, the amount of waste, the CCA, CAP and UCA indices are very difficult to implement.

Therefore, these indices are still far removed from business discussions, and it is difficult to explain their real meaning for business management.

Indicator 8 - Indicators of environmental conditions -

AIR - contaminants, shows soot from diesel engines that have never been measured;

Noise - there are no ongoing measures, only those in the PPRA, which is already four years old;

Odors - there are odors due to the consumption of diesel oil and in the toilets; Water - contaminants, presents analysis based on CONAMA resolution 20, from the EIA - RIMA carried out.

Soil - there are areas prone to erosion, although there are no outbreaks;

Flora - the EIA-RIMA presents a description and quantification of the flora in the mining area;

Fauna - same as above;

Men - there are no health indicators in the company, primarily due to the non-implementation of the PPRA; they participate in vaccination campaigns only by informing their employees of the existence of the government campaign.

There is no archived data on diseases linked to public health.

It has no concern for the housing of its employees, nor for the construction quality of the company's buildings.

The same goes for the water distribution and sewage treatment service, which uses black pits to collect liquid effluents and the water comes from a deep tube well, a work that has no technical manager, no maintenance and no use permit.

There is no health center in the vicinity of the extraction site, only a first-aid box in the company's administration office and on the boats (dredgers) that are used for extraction in the river.

There is no school nearby and no agreement for employees to study.

4.2.2 Mining in the municipality of Icaraima, PR.

This sand mining company is more organized and the geologist in charge of it received the questions in advance, read them before the interview and expressed his knowledge of some of the criteria discussed during the conversation. It's a sand extraction company that has had some problems with the state environmental agency, so it has experience in dealing with environmental legislation and its requirements, as well as some accumulated discussion.

Indicator 1 - Environmental policy: There is no written policy, there is only a discussion about the importance of regularizing the company, given the problems it has already faced, considering the reduction in production caused by the stoppage and not the larger environmental issue.

Indicator 2 - Environmental Management System (EMS): There is no EMS, the responsibility lies with the manager (who lives in the mining company) and who can be considered to be a person who is poorly prepared for the job, as he has little technical knowledge of the work he does.

Indicator 3 - Risk Management System (SGR): It has a PPRA, but it doesn't know who did it (engineer or safety technician), it's not in the company's office, nor is it in place, and there has been no safety training for employees.

Indicator 4 - Legal Management System - LMS : In terms of legal compliance, there is no ordering of legal proceedings and fines.

There is no one responsible for organizing the information, nor can they say what the fines are and their monetary value, much less the notifications or communications from the state environmental agency. Therefore, there is no report from or to the company's head office (which is located in another city) that can administratively record this data.

Indicator 5 - Product management: There are no innovations in the extraction process or new products. However, it stands out for having implemented an environmental recovery plan, highly praised by state and municipal environmental agencies. It has no cooperation programs between companies, although its owner belongs to an association of evangelical businessmen, the Full Gospel Businessmen's Association - ADHONEP .

No proactive environmental operational strategies.

It does not use the community for the company's environmental and social projects.

It also does not present a critical evaluation of any program it has participated in or implemented.

Indicator 6 - Environmental education projects : There are no waste reduction programs, although some recycling already exists (using sheets of paper for drafts that have already been used on the other side and collecting newspapers to deliver to nursing homes; there is a small electricity saving program.

It is not involved in other social programs in the community.

Indicator 7 - Financial indicators : For an analysis of its financial performance, information is very difficult to obtain, as its directors and technicians do not like to comment on its turnover in cash and its relationship with the investments made.

The same goes for environmental spending, which has not been separated from the totals spent and in its balance sheet.

They couldn't say how much electricity is used.

They don't have an organized record of the total amount of fuel consumed, in liters, over any given period. They only knew that it was around 700 liters a week. They can't say for sure how much they produce and how much they consume; they don't know how much waste they produce. They are therefore unable to determine the production losses caused by environmental problems.

Therefore, it cannot cope with the CCA, CAP and UCA calculations, thus jeopardizing the other proposed indices (environmental projects and fines, environmental investments and totals, electricity and production, fuel and production, fleet expenses and production).

Indicator 8 - Indicators of environmental conditions: Air - no measurements of air contamination.

Noise - there were only measurements from the PPRA, which date back to 1999. Noting that there were no odors in the bathrooms.

Water - presented an analysis based on CONAMA 20, present in the Environmental Control Plan - PCA, in 1998 and no longer took any samples.

Soil - all hotspots and areas susceptible to possible erosion were treated by the PCA, in the recovery required by the Paranà Environmental Institute.

Flora - presented a descriptive study and maps of occurrence in the PCA.

As for the man - he doesn't bother to take part in health campaigns, nor to list the illnesses of his employees; he doesn't describe or worry about the housing of his employees, nor the public health of the area around the enterprise.

It doesn't have a health center or a school for its employees or the surrounding inhabitants.

4.3 Discussion of the data collected and other considerations.

4.3.1 Introduction.

Using the concept developed by WINSSEMIUS (op. cit.), extraction in the municipality of Guairà can be classified as a functionally oriented company, or rather, its staff do not have the specializations needed to deal with environmental problems, but focus solely on the running of the company.

Mining in the municipality of Icaraima also has the same characteristic: it is functionally oriented, although it does have some experience in dealing with the environment. This was acquired as a result of the licensing process it underwent and the obligation to present a project to recover the area for sand deposits, and the environmental agency's supervision of its execution. These two companies keep their production focus traditionally oriented and disconnected from environmental issues. They are companies that only comply with current legislation and are only concerned with implementing the most efficient solutions from a cost point of view.

KINLAW (op. cit.) also sees the two extractors as reactive companies. The two mining companies only react to the nuances and pressures of the market, and do not display policies that are compatible with the new market that is emerging in the world.

4.3.2 Indicator analysis and associated practice

Indicator 1, the environmental policy, as a basic indicator for any analysis of environmental performance, occupies a prominent place in demonstrating the type of concern that the organization has for the environment (in addition to occupational safety and health issues and social issues) for a performance of continuous improvement.

The discussion of this indicator seeks to integrate management systems, and is concerned with the close adherence of processes and practices between the ISO 9000, ISO 14000 and OHSAS 18000 standards and the day-to-day activities of mineral extraction.

Thus, the lack of a clear environmental policy shows that the practice is dissociated from the processes and actions recommended for environmental management, contributing to the other functions of the organizations analyzed prioritizing and valuing the daily routine.

These actions to add value to the daily routine leave no room for environmental initiatives that could add value to their product. Thus, at the mining company in Icaraima, the employee assigned to take care of these actions has not grasped the strategic significance of the environmental issue and is an employee who has not been properly trained. As far as the second indicator is concerned, the two companies obviously do not have an EMS in place, because without a clearly presented environmental policy, an EMS cannot be set up and there is no determination of significant environmental aspects for monitoring and measurement for continuous improvement.

According to the ISO 14000 standard, the plans that make up the programs must define the responsibilities, interfaces and relationships, deadlines, means and resources needed to make them effective. None of the aforementioned requirements were observed, namely the responsibilities, interfaces, relationships, deadlines, means and resources needed to make environmental management effective.

As for the third indicator, the aim is to diagnose concern about the other side of the environment, occupational health and safety. In the same way, as in the previous case, the RMS should show the main risks, both environmental and occupational, as well as emergency situations and the measures taken to deal with them.

The data analyzed showed only the legal documents. In one case, the documents are known (Guaira), in another (Icaraima) they are unknown. In both cases, the extraction sites have no mechanisms for controlling risks, only their identification, based on the PPRA.

The two environmental risk prevention plans are not in place, there is no monitoring, no critical analysis and no regular audits. This means that organizations are not concerned about their internal customers, their main collaborators.

With regard to indicator 4, legal compliance, none of the two companies showed the necessary concern for this organizational system. All two employees who answered the questions had to call or ask permission from management for a telephone consultation with the lawyer in charge. The two legal professionals couldn't answer immediately, as they didn't understand the LMS mechanism.

They did not present a consistent report with an administrative record of the progress of the environmental lawsuits against the company, nor did they aim to demonstrate a reduction in the costs and expenses of fines and notices.

The company's legal management implies a commitment to meeting the minimum requirements of legislation, as well as a commitment to continuous improvement. The legal level is the beginning of the company's environmental concern.

As a criterion, product management (indicator 5), seeks to show concern for the product (sand, in our case) and innovations in its process.

Only one, Guairà, has shown concern for improving the product, trying to implement a differentiation of granulometry to increase its added value.

There is no association that defends the business interests of the miners, and competition is extreme.

Their operational strategies have remained the same for decades, with no changes that could reflect any alteration in the quality of the product.

They don't use the community for marketing actions, nor do they present a critical evaluation of any social program in which they have participated or implemented.

Indicator 6: both sand extraction companies don't have environmental education projects, they don't use recycling, for example, of the used oil that is removed from the maintenance trucks.

The Icaraima quarry is experimenting with some practices (which have already been mentioned), while the Guairà quarry is well behind schedule and no practices can be detected in this area.

Financial performance, as the seventh indicator, proved to be difficult to obtain, with no concrete data.

The figures varied greatly and those who responded did not confirm any of the data that was presented (electricity, environmental investments, fuel costs, quantities of waste produced, related to sand production, among other figures).

The numbers used to calculate the CCA, CAP and UCA were difficult for the technicians and accountants to understand.

They don't publish balance sheets or even know what a social balance sheet means. In their chart of accounts, they do not distinguish items that relate to the concepts of environmental assets and liabilities.

During the interviews, the association made by LEMOS et all (1986), who measured sand production in the municipality of Juquià -SP, and observed that only 65% of actual production was officially declared, became very clear. This leads us to think that neglecting financial performance is just a smokescreen to prevent us from seeing the company's real situation.

The last indicator, environmental performance, was also incomplete.

The first (Guaira) submitted an EIA - RIMA as part of its environmental licensing, while the other (Icaraima) also submitted a PRAD as part of its environmental licensing with the IAP.

The PRAD only focused its observations and actions on the recovery and maintenance of the surrounding landscape, and did not present measures of the probable

air emissions, noise and odors, as well as physico-chemical analyses of the water and observations on the fauna. Although I mapped the points that showed erosion, the density and dominance of the flora were treated theoretically.

The EIA-RIMA presented adds very little to the observations of regional parameters, uses the same data as the PPRA when dealing with noise and emissions, does not present a physical-chemical analysis of the water and also deals with flora and fauna in a theoretical way in relation to their density and dominance.

At the end, it presents a series of actions aimed at mitigating the impacts observed, without providing a measure of the performance of these actions.

4.4 Discussion of results

And finally, what can be observed is that these two companies have failed to maintain solid controls for the management of their businesses, making the environmental discussion an imperceptible issue for their owners, who find themselves in a difficult managerial situation.

Environmental management and its performance evaluation, which has been recommended as a validation of this study, could be an aid to the organization of these companies.

4.4.1 Regarding the environmental performance assessment proposal

At the beginning of this work, we tried to demonstrate that the development of environmental performance indicators favors better business decisions, being significant in achieving continuous improvement and in the evolution of business strategy.

The company is forced by the competition to create performance indicators that are only financial in nature, and is led to implement strategies to compete in the market, only taking into account the cost aspect. It analyzes its production process, organizational learning, both in terms of operation (production process) and human relations, developing an understanding of environmental issues that is also based on costs, from the point of view of expenses.

The dimensions of quality, protection of the natural environment, employee health and safety are not seen as factors in the company's success.

They fail to understand strategically that indicators for controlling their processes, not only environmental changes, but also occupational health and safety and the quality of their production process, reduce costs, help to recognize critical points in production, and define responsibilities and standards to follow.

Applying ABREU's (2001) reasoning, we have companies with weak environmental conduct, where qualitative indicators are developed to the detriment of quantitative ones. These behaviors can be identified through the answers given to the first questions (environmental policy, EMS, RMS and LMS). Their answers seem to strengthen the quantitative aspects of environmental performance evaluation.

4.4.2 For the sand extraction industry

As a whole, the sand extraction companies in the region can benefit from the indicators developed here, as the ADA will create an environment favorable to regional competition, adding value to the product and the image of the sand extraction sector. With the application of this model, companies that used to be reactive can now compare their individual performance on four levels.

This gives :

1. companies at the reactivity level;

2. companies starting to implement ADA;

3. companies in the ADA consolidation stage (up to the third year of implementation);

4. proactive companies (after the fourth year of implementation).

On the other hand, for environmental agencies, it could represent a way of evaluating and achieving compliance, both in relation to the provisions of the legislation and the inspection parameters used by these agencies.

The mining region where the questionnaire was carried out is, by law, an Environmental Protection Area (APA), which allows for various types of land use, as long as they are regulated by technical criteria. In this way, the ADA could prove to be an excellent regulatory guide for the discussion of land use criteria in the region.

5. Conclusions and recommendations

5.1 Introduction

This work concludes with the valid use of environmental performance assessment as an environmental management tool for a sand extraction company using the dredging method. The indicators were obtained thanks to a comparative study of the environmental impacts of this method of extraction, with the bibliography found on ADA, fulfilling its general objective.

The set of indicators used to establish the ADA proved to be robust for assessing the environmental performance of sand ports, since it was easy to apply and understand, requiring some time to adapt and discuss the technical nomenclature used.

These indicators were obtained through a conceptual foundation, providing a comprehensive view of the environmental impact of sand extraction, using the dredging method, on riverbeds.

At the end of the previous chapter, the environmental performance of companies can be classified, which leads us to some reflections:

- The two mines did not have any information to answer the questionnaire;
- The companies did not have reliable data on the qualifying indicators (environmental policy, EMS, RMS, LMS and product stewardship);
- The failure to present qualifiable data hindered the presentation of quantifiable indicators, because if there is no environmental policy, there is no environmental practice, and without practice there is no data to be collected;
- This data should primarily come from the efficient application of the first part of the indicators, the non-quantifiable indicators (environmental policy, environmental management system, risk management system, legal management system and product stewardship);
- The second part, the quantifiable indicators, was hampered by the failure to present the aforementioned indicators.

Therefore, the questions presented as indicators proved to be applicable for qualifying the environmental performance of companies, establishing in their application a first classification of environmental performance, that of reactive.

On the other hand, it has also proved to be an effective way of initiating further discussion of environmental indicators for comparison between companies (*benchmarking*).

The environmental conduct of these companies can be identified as qualitative and quantitative, leading to the perception that, in theory, the qualitative indicators are more associated with environmental conduct and the quantitative indicators are more related to performance.

On the other hand, also in theory, it can be seen that the qualitative indicators precede the quantitative ones, since the two companies in question are first under pressure from external forces, which generates within them a determination to overcome this pressure. This gave rise to environmental policy guidelines and quantification through environmental condition indicators.

5.2 Conclusions

As this is a research project that seeks to establish environmental performance indicators for sand extraction using the dredging method in riverbeds, it was possible to identify a reactive level of performance and was unable to develop other levels of performance due to the lack of information from the business community, local managers and technicians involved. Thus, the series of indicators proposed allows us to measure the environmental performance of these mines. They can even be divided into four classes of environmental performance, namely :

a) reactive;

b) in the early stages of implementation;

c) after three years of implementation, with its evaluation and performance system already established;

d) proactive.

5.3 Recommendations for future studies

As a continuation of the studies, it will be necessary to apply the same questionnaire to other mines of the same type (sand extraction by dredging, in riverbeds), but in other locations to see if their behavior is the same.

Secondly, in other extraction models (pit and dry sand extraction, among others) and applied to mining of other minerals (clay, basalt, among others) to see if the results reported here are still valid as a measure of environmental performance for small mining operations.

In addition, it should aim to draw up training programs for the business community, as well as lectures for employees and, based on the results obtained, seek a better way of managing the environmental issue and its corollaries, overcoming the current state of affairs.

Bibliographical references

BRAZILIAN ASSOCIATION OF TECHNICAL STANDARDS - ABNT Characterization of pollutant loads in mining (project 1:602.02 - 006) Procedure - 1st draft of standards, Oct, 1996.

ASSOCIAÇÃO BRASILEIRA DE NORMAS TÉCNICAS - ABNT Environmental Management System - Specifications and guidelines for use: NBR/ISO 14.001. São Paulo: ABNT, October 1996.

ALMEIDA, Josimar Ribeiro; MELLO, Clàudio dos S.; CAVALCANTI, Yara, Gestâo ambiental : planejamento, avaliaçâo, implantaçâo, operaçâo e verificçâo. Rio de Janeiro: Thex, 2000.

ANDRADE, Rui Otâvio Bernardes de; TACHIZAWA, Takeshy e CARVALHO, Ana Barreiros de. Gestâo ambiental - enfoque estratégico aplicado ao desenvolvimento sustentâvel, Sâo Paulo:Makron, 2000.

BRAZIL, Law No. 6.514, of 22/12/1977, amends Chapter V of Title II of the Consolidation of Labor Laws on occupational safety and medicine. Manuais de legislaçâo Atlas, Sâo Paulo:Atlas, 1999.

BRUSCHI, Denise Marilia e PEIXOTO, Mônica Campolina Diniz, Extraction of sand, gravel and clay - Techniques and environmental control, Belo Horizonte: State Foundation for the Environment, in: Manual de Saneamento e Proteçâo Ambiental para os Municipios, V.4, 1997.

CALLENBACH, Ernest; CAPRA, Fritjof; GOLDMAN, Lenore; LUTZ, Rudiger; MARBURG, Sandra; Gestão ecológico (ecomanagement), Gua do Instituto

Elmwood de auditoria ecològica e negócios sustentàveis, Sâo Paulo: Cultrix Amanà, 9ed. 1999.

CARO, Nauri Miguel Heriberto and RADOS, Gregório Jean Varvakis, Performance measures as a basis for continuous process improvement: the case of the Fundaçâo de Amparo à pesquisa e Extensâo Universitària, dissertation, Florianópolis: FAPEU, 1998.

CARVALHO, Alexandre Bruno Moreno de, Focus on quality, environment, health and safety, in: Revista Banas Ambiental, pp. 46 to 52, December/2000.

CARVALHO, L. Nelson, MATOS, Emanuel Rodrigues de and MORAES, Romildo de Oliveira Environmental Accounting, handout, n.d.

CARVALHO FILHO, Elcio Teixeira de, Prevençâo de perdas, handout, 1998.CASTRO, Armando G. B. de, Açôes corretivas e preventivas para a harmonizaçâo das actividades de mineraçâo com a preservação do meio, em Nova Lima, MG. In : Course on pollution control in mining: some aspects, vol. 2, DNPM, Brasilia, 1986, p. 09-24.

CAVALIERI, Ângelo Salvatore. Avaliação de desempenho ambiental : uma proposta de relatório e indicadores de desempenho ambiental para a divulgação dos resultados das empresas, dissertation, Rio de Janeiro: UFRJ, 1997.

COHEN, Mark A., FENN, Scott A. and KONAR, Shameek. Environmental and financial performance: Are they related?, in : http://www.vanderbilt.edu/VCEMS/papers/envperf.html, consulted on 31/07/2000.

CONAMA, Resoluçâo n.° 001, de 23 de janeiro de 1986, dispôem sobre a elaboração de definições, responsabilidades, critérios básicos e as diretrizes gerais para a implementação da Avaliaçâo de Impacto Ambiental, como um dos instrumentos da Politica Nacional do Meio Ambiente, in : Diàrio Oficial da Uniâo, de 17/02/86, pg. 2548.

CONAMA, Resoluçâo n°020, de 18/06/1986, dispôem sobre a classificaçâo das águas doces, salobras e salinas do territòrio nacional, in: Coletânea de Legislaçâo Ambiental, 1995.

CONAMA, Resoluçâo n.° 010, de 14 de dezembro de 1988, dispôem sobre o zoneamento ecológico-econòmico das APA's e da outras providências, in: Coletânea de Legislaçâo Ambiental, 1995.

CREA - PR, Regional Council of Engineering and Architecture, Act No. 45/97 - Classifies sand mining according to production limits, 16/09/1997.

CZAJA, Mauricio Camargo, Implementation requires proactive management, in: Banas ambiental, revista, no 12, pág. 46 to 50, Sâo Paulo: Banas, 2001.

D'AVIGNON, Alexandre. ISO 14000 environmental standards: how they can influence your company, Rio de Janeiro: CNI, 1995.

DEMAJOROVIC, Jacques and SANCHES, Carmen Silvia. Education and environmental indicators: perspectives for organizations. V Encontro Nacional Empresarial e Meio Ambiente, proceedings, November 17-19, 1999, FEA/USP.

De SIMONE, Livio D. and POPOFF, Frank. Ecoefficiency, the business link to sustainable development, Word Business Council for Sustainable development:MIT, 1997.

DeCICCO, Francesco. Introduction to management systems (quality, environment, safety and health), workshop, Sâo Paulo: Risk tecnologia, 1995.

DOWN, C.G. ,STOCKS, J. Environmental Impact of mining, reprinted, London: Aplied Science Published, 1978.

FERREIA, Araceli de Souza. Contabilidade ambiental Custos ambientais - uma visâo de sistema de informaçâo, in: http://www.wwiuma.org.br/contb_ambiental_af.htm, consulted on 16/02/2002.

GALVÀO FILHO, Joâo Baptista.8888 Gestâo de risco ambiental, in: Banas ambiental, revista, no 12, pág. 32 to 41, Sâo Paulo:Banas, 2001.

FUNDACENTRO. Course in occupational engineering, several volumes, 1981.

GAZETA MERCANTIL, Environmental Management - a company commitment, 8 issues, 1996.

GIL, Antonio C. Métodos e técnicas de pesquisa social. 4 ed. Sâo Paulo: Atlas, 1995.

GREGORIS, Joâo Carlos D.M.C. Course on qualitative risk analysis techniques, handout, 1994.

GUERRA, Antônio José Teixeira e CUNHA, Sandra Baptista da Cunha (org.) Geomorfologia e meio ambiente, Rio de Janeiro: Bertrand Brasil, 1996.

HARRINGTON, H. James and KNIGHT, Alan. Implementing ISO 14000 - How to update your environmental management system effectively. São Paulo: Atlas, 2001.

HRONEC, S. M. Vital Signs, São Paulo: Makron, 1994.

JOHR, Hans, O verde é negócio, São Paulo: Saraiva, 1994.

JUCHEN, Peno Ari, Introduçâo à gestâo, auditoria e balanço ambiental para empresas.Curitiba, xerox, 1995.

JUCHEN, Peno Ari (coord.), GIMA, guia de indicadores e métodos ambientais, Curitiba: IAP/GTZ. 1992.

KAPLAN, Robert S. and NORTON, David P. Strategy in action - Balanced Scorecard. Rio de Janeiro: Campus, 1997.

KINLAW, Dennis C. The competitive and ecological company - strategies and tools for conscious, responsible and profitable management. São Paulo: Makron, 1997.

KONAR, Shameek and COHEN, Mark A., Does the market value environmental performance in: http://www.vanderbilt.edu/VCEMS/papers/mkt value.html.

KOPEZINSKI, Isaac, Mineraçâo x meio ambiente - considerações legais, principais impactos ambientais e seus processos modificadores. Porto Alegre : Ed. Universidade/UFRGS, 2000.

LEMOS, Antônio Carlos P. N. ; CAMPOS, Heraldo C. N. S.; BITTAR, Omar Yazbek and MONFORT, Roberto Antônio. Mining planning and management: a proposal

for municipal participation in mining planning and development. Bol. Téc. SUDELPA, 1986.

MACHADO, R. G. (coordinator) Environmental Control Plan (F. Andreis e Cia Ltda, Porto Figueira, Vila Alta), report, 1998.

MACHADO, R. G. (coordinator) , Environmental Control and Recovery Plan (Porto Camargo - Lira Mineraçâo Ltda, Icaraima, PR), report, 1998 a.

MACHADO, R. G. (coordinator) Environmental impact assessment (Andreis Group, D'agostini Group, Pawlowski Group and Requiâo Group), report, 2000.

MARTINS, Eliseu e de LUCA, Màrcia Mendes Ecology via accounting, in: Inter-American Accounting Conference, Proceedings, Santo Domingo, 1993.

MEDEIROS, Maria Lùcia M. B. Considerações sobre avaliaçâo de impactos ambientais em ecossitemas terrestres in: MAIA - Manual de avaliaçâo de impactos ambientais.Curitiba:SEMA/IAP/GTZ, 1995.

METCALF, Kim R.; WILLIAMS, Phillip L. ; MINTER, J. Robert; HOBSON, Chris M., An assessment of corporate environmental programs and their performance measurement systems, in: Journal of Environmental Health, vol. 58, n.2, p.9 to 17, 1995.

METCALF, Kim R. ; WILLIAMS, Phillip L.; MINTER, J. Robert; HOBSON, Chris M. , Environmental Performance Indicators for enhancing environmental management, in: Total Quality Environmental Management, Vol. 5, n.4, p. 7 to 11, 1996.

METCALF, Kim R., WOODALL,W. Robert, HOBSON, Chris M. and WILLIANS, Philip L. Environmental Performance measurement : A case study, in: Environmental Quality

Management, vol. 5, n.5, p. 27 to 37, 1996a.

MINEROPAR (Minerals of Paranà) Mining and the Environment, Curitiba, Government of the State of Paranà, 1991.

GERMANY'S MINISTRY OF THE ENVIRONMENT (Federal Environmental Agency). A guyide to corporate environmental indicators, 1997.

MOURA, Luiz Antônio Abdalla de. Quality and environmental management - suggestions for implementing ISO 14.000 standards in companies. Sâo PauloJuarez de Oliveira, 2002.

MULLER, A. C., Hidrelétricas, meio ambiente e desenvolvimento. Sâo Paulo: Makron, 1995.

PARIZOTTO, José Antônio.O gerenciamento ambiental:estudo de caso de cinco empresas de mineraçâo no Brasil. Rio de Janeiro:CNPq/CETEM, 1995.

PORTO, Clàudio, Uma introduçâo ao planejamento estratégico, in: Boletim tècnico do Senac, http://www.senac.br/boletim/boltec35.htm, consulted on 20/03/2001.

RADOS, Gregório J. V. ; DIAS, Paulo M.; CARO NAURI, Miguel; NERES, Wudson Anthony, Gerenciamento de processos, master's course in production engineering, environmental quality management, UFSC, PPGEP, handout, 2000.

REIS, Mauricio J. L. ISO 14.000 - Environmental management, a new challenge for competitiveness. Sâo Paulo: Qualitymark, 1995.

SANCHEZ, Luiz Enrique. Desengenharia - O passivo ambiental na desativação de empreendimentos industriais, Sâo Paulo:Universidade de Sâo Paulo, 2001.

SEBRAE, A questão ambiental - o que todo empresârio precisa saber, Brasilia: SEBRAE, 1996.

SILVA, H. V., Plano de Recuperação Ambiental - "Porto de Areia : PACAJAS", document used by the IAP (Environmental Institute of Paranà) to issue the operating license, 1992.

SILVA, H. V., Changes in the quality of surface water due to sand mining in the municipality of Itaquaquecetuba - SP. In : Boletim de Geografia - UEM, n.01, Maringà, 1996, p. 01 - 06.

SILVA, H. V.; Mining and environmental planning: an experience in the greater São Paulo region and a proposal for methodology. Rev. UNIMAR - UEM, n.19, vol. 4, Maringà:EDUEM, 1997, p. 1149 to 1162.

SILVA, H. V. (coord.) Avaliação de Impacto Ambiental EIA-RIMA (Mineraçâo Floresta -

Guaira, PR) report, 2000.

SILVA, H. V. As metodologias de anâlise de risco, monografia de especialização em engenharia de segurança, Maringà:UEM, 2000a.

SILVA, H. V. and MACHADO, Joâo A. de P., Environmental Recovery Plan - Olaria : Gilmar Peterman, report, 1992.

SILVA, H. V. and RADOS, G. Varvakis, O Gerenciamento de processos como um método de avaliaçâo de impactos ambientais, Proceedings of the VIII Encontro da Seçâo

IAIA (International Association for Impact Assessment), November 23-26, 2000, Palhoça, SC, unpublished.

SILVA, Valéria de Càssia, Estudo das condiçôes de sanamento do Balneàrio de Canasvieiras, visando ao gerenciamento ambiental, dissertation, PPGEP, UFSC:Florianópolis, 2000.

SIMM, Sandra. Environmental awareness and management as strategy shapers. Londrina: UNOPAR, xerocopy, 2000.

VALLE, Cyro Eyer. Environmental quality - the challenge of being competitive while protecting the environment. Sâo Paulo: Pioneira, 1995.

VITERBO JÛNIOR, Ênio Sistema Integrado de Gestão Ambiental : como implementar um sistema de gestão que atende à norma ISO 14000, a partir de um sistema baseado na norma ISO 9000, Sâo Paulo:Aquariana, 1998.

WINSSEMIUS, Pieter. In search of a business agenda for the transformation of environmental performance, insert in Novo Ambiente magazine, October 1997.

Annex

Annex 1

The following indicators are described below in the form of questions and/or numerical data collection:

1. Environmental policy

Is there an environmental policy, does it have goals, a scope, an introduction date, a validity period or a review?

Is there explicit reference to the issue of continuous environmental improvement?

Yes / No / In progress

2. Environmental management system

Structured management, including organizational responsibility with action and training plans ?

Yes/ No/ In progress

3. Risk management system

Does it present plans describing accidents and the PPRA environmental risk prevention plan (Law 6514 / 77), describing risk prevention programs? Yes/ No/ In progress

4. Legal compliance

How many and what are the court cases involving environmental issues?

What is the number of fines and their total monetary value? n°

value ___________

Does it present a report of conformities and non-conformities ?

Yes /No In progress

What is the number of communications or notifications from the environmental agency?

N°

5. Product management

What innovations (eco-efficiency, environmental design, new products and processes)?

- Number of projects ______________________

What cooperation programs exist?

- Number of programs
- Number of operational strategies on environmental quality
- Number of people from the community involved

Does it present a critical evaluation of these cooperation programs?

Yes/no/ongoing

Are there any waste reduction programs?

Gaseous emissions ?

Net emissions ?

Material recycling and the associated cost ?

Energy efficiency program ?

- Yes/no/ongoing

6. Other programs ? (environmental education, competitions in schools, cultural sponsorships)

- Yes/no/ongoing
- Number of participants

How many people participate in environmental projects (for the government or NGOs)?

7. Financial performance

Total currency turnover and total investments and their ratio, measured in the fiscal year$/$, per month and accumulated in the fiscal year

- Total spent on environmental programs, total investments and their ratio, measured in the fiscal year $/$, per month and accumulated in the fiscal year
- Total spent on electricity, related to the amount of sand produced, measured in the fiscal year $/ton, per month and accumulated in the fiscal year
- Total spent on fuel (petroleum-based) in relation to the amount of sand produced, grouped in the fiscal year $/ton, per month and accumulated in the fiscal year

• Total fuel consumption in the fleet in relation to the amount of sand produced, grouped in the fiscal year liters/ton, per month and accumulated in the fiscal year

• 8. Environmental development

Air

Contaminants (CONAMA resolution)

Noise (expressed in the PPRA)

Odors (present or not)

Water

Contaminants (CONAMA resolution)

Erosion spots (present or not)

Areas prone to erosion (present or not)

Soils

Quality (mapping ?)

• Which species per square meter

Quantity (is there monitoring?)

• number per square meter

Quality (mapping ?)

• description of species

• number of species around the mine and in the mining area quantity in the entire mining area (is there monitoring?)

Man

Health indicators

Vaccinations

• participates in vaccination campaigns (yes/no/ongoing)

• presence of public health-related illnesses in the fiscal year (number and type)

Housing indicators

Quality

-Water distribution service (yes/no/ongoing)

- and sewage treatment (yes/no/ongoing)
- Health center nearby (yes/no/ongoing)
- School nearby (yes/no/ongoing)

MIX
Papier aus verantwortungsvollen Quellen
Paper from responsible sources
FSC® C105338

Printed by Books on Demand GmbH, Norderstedt / Germany